ShapingourEnvironments

LEVEL3GEOGRAPHY

Operation of Tourism Development in a New Zealand Environment

Elliot Moka

Operation of Tourism Development in a New Zealand Environment
1st Edition
Elliot Moka

Cover design: Cheryl Smith, Macarn Design
Text design: Cheryl Smith, Macarn Design
Production controller: Siew Han Ong

Any URLs contained in this publication were checked for currency during the production process. Note, however, that the publisher cannot vouch for the ongoing currency of URLs.

Acknowledgements
The author and publisher wish to thank the following people and organisations for the resources in this textbook:
Front cover: Tourists on Redwoods Treewalk at The Redwoods (Whakarewarewa Forest), Rotorua, North Island, New Zealand. David Wall / Alamy Stock Photo.
Back cover: View of Rotorua from above Te Puia. Bailey Beaufill / Shutterstock.
Shutterstock for the images on pages 4, 5, 7, 8, 16, 17, 18, 20, 21, 22, 23, 24, 27, 31, 38, 39, 40, 41, 45, 46, 50, 57, 63.
Destination Rotorua for the images on pages 31 (Mitai Maori Village); 43 (top) Velocity Valley, (centre and lower) Skyline Rotorua; 44 (top) Tamaki Waka, (DR), (lower) Rotorua Marathon, Up&Up Photography & Film; 45 (top) Redwoods Treewalk, (centre) Rotorua Canopy Tours; 46, Rotorua Energy Events Centre; 63 Polynesian Spa; 65 Rotorua Duck Tours.
Alexander Turnbull Library for the images on page 35, Willis, Archibald Duddington (Firm): Tarawera eruption, N. Z. 1886. *A Happy New Year.* Wanganui, A.D. Willis, [ca. 1886]; page 36, (top right) New Zealand Railways Publicity Branch: *Rotorua, nature's cure. Thermal waters, health and recreation. Best reached by rail, New Zealand.* Issued by the New Zealand Railways Publicity Branch [ca. 1932]; (top left) View of the thermal area at Tikitere showing a group of visitors being guided around the area. Photograph taken by William Price in 1908; (bottom) New Zealand Railways Poster — Tourist Excursion Tickets First Class, 1 November 1888 – 31 March 1889; page 40, (top) Rotorua Bus Company Ltd. Short trips in Rotorua, New Zealand's scenic paradise, (centre) Rotorua Airport and Rotokawa Bay, 1963, Aerial photograph taken by Whites Aviation.
Other images: page 34, (top) Robert Wynyard, *Pah on Rotorua Lake with hot springs*, 1849, (lower) John Backhouse (1845–1908), *The White Terraces*, 1886; page 38, *White Terraces, New Zealand*, Charles Bloomfield, 1884.

© 2021 Cengage Learning Australia Pty Limited

For product information and technology assistance,
in Australia call **1300 790 853**;
in New Zealand call **0800 449 725**

For permission to use material from this text or product, please email **aust.permissions@cengage.com**

National Library of New Zealand Cataloguing-in-Publication Data
A catalogue record for this book is available from the National Library of New Zealand.

9780170446914

Cengage Learning Australia
Level 7, 80 Dorcas Street
South Melbourne, Victoria Australia 3205

Cengage Learning New Zealand
Unit 4B Rosedale Office Park
331 Rosedale Road, Albany, North Shore 0632, NZ

For learning solutions, visit **cengage.co.nz**

Printed in Singapore by C.O.S. Printers Pte Ltd.
2 3 4 5 6 7 25 24 23 22

Contents

1 Introduction to tourism 4

2 Cultural processes 7

3 The tourism development process 14

4 Tourism development in Rotorua 29

5 Temporal variations 33

6 Spatial variations 52

7 Tourism impacts 62

Answers 68

Introduction to tourism

What is tourism?

Tourism is people travelling away from their homes to places of interest. This travel could be for sightseeing or leisure like a beach holiday in Hawaii, for business activities like meetings or conferences, or for visiting friends or family for events like weddings or celebrations.

Tourism activities are important because of the economic benefits for people all over the world. When tourists travel to new locations, they spend money, which goes into the local economy and provides jobs and incomes. For example, tourism-related activities typically contribute around 20 percent of New Zealand's export earnings from the spending of international tourists in our country. Domestic tourists spend even more.

Tourism activities draw visitors from within countries, but also from across the world, making tourism a global activity. The impacts on people and places can be clearly seen in times of economic downturn like the global financial crisis, which started in 2008, and the global Covid-19 pandemic, which started in 2020. In these times, visitor numbers and spending decreased dramatically. This has created negative impacts for people and economies due to job losses and lost incomes.

It is important to understand tourism and tourism-related systems and processes as a whole so we can better understand the impacts and plan for future changes.

Although tourism activities occur all over the world, there are differences in visitor numbers between countries and locations.

📍 Tasks

1 These are the world's top 10 tourist destinations by number of international tourist arrivals.

Ranking	Country
1	France
2	Spain
3	United States
4	China
5	Italy
6	Turkey
7	Mexico
8	Germany
9	Thailand
10	United Kingdom

a Locate and label these countries on the map below. Give the map a title.

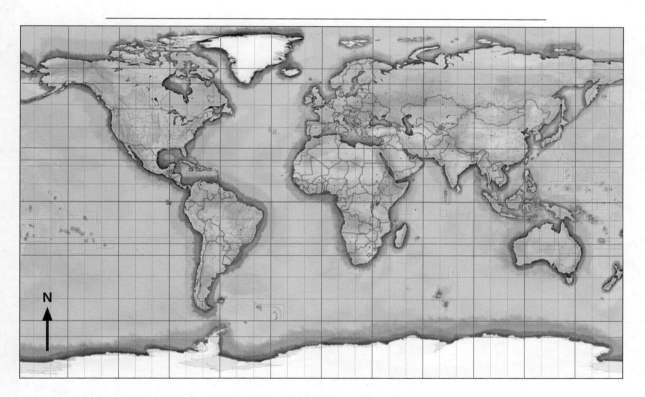

b Describe the pattern of countries seen in the map.

_____

_____

_____

_____

_____

2 a For the countries listed below, complete the table to show the attractions that people would
      want to visit and why.

Country	Natural attractions	Cultural attractions	Reasons why people would visit these attractions
France			
United States			
China			

   b Explain why people would want to visit these countries.

   _____

   _____

   _____

   _____

   _____

Cultural processes

Processes are at the core of geography because they are a sequence of actions that bring change to people, places and environments.

Cultural processes are processes started by people, usually for social and/or economic reasons like migration and industrialisation.

Human involvement makes cultural processes different from natural processes. Natural processes occur on their own without the input of people. Examples of natural processes are tectonic processes and climate processes.

Cultural processes modify both natural processes and environments, as people modify natural systems to suit their needs. This can be seen in the processes of mining and urbanisation, for example.

Tasks

1   Match each process from the list below to the related image and categorise each as either a natural or a cultural process.

Word list: Agriculture, Horticulture, Forestry, Vegetation succession, Aeolian deposition, Electricity generation.

Image	Process	Process type: Natural or Cultural

Image	Process	Process type: Natural or Cultural

2 Explain why some processes are cultural when they involve inputs from the natural environment.

_____

_____

_____

_____

_____

_____

Elements and interaction

Elements are key features with their own unique characteristics that are part of a cultural process. These include tourists, attractions, and accommodation types.

Interaction is an important key geographic concept for this topic as it relates to how elements affect each other in an environment.

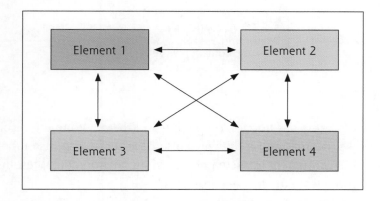

The diagram above is a simple way to show the interactions between elements in an environment. The more elements there are, the greater the number of interactions.

The diagram below uses the examples of attractions, tourists, and accommodation. Different types of attraction might complement each other or compete against each other for tourist numbers and spending. Tourist preferences and needs will determine what they will spend their money on when given choices between attraction and accommodation types. The way the elements interact with each other leads to the operation of the cultural process.

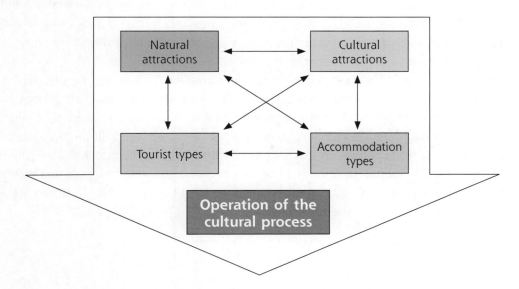

The operation of a cultural process

It is easier to analyse cultural processes if we can break them down and examine the key parts as a system. A system involves inputs, processes, outputs, and feedback. This can be seen in the diagram below.

- Inputs are elements added into a system. This could be money invested.
- Processes are the actions that bring **change**, like tourist visits and spending.
- Outputs are the end results after a process has occurred, like profits.
- Feedback is about returning outputs back into the system as inputs. Feedback could be positive or negative.

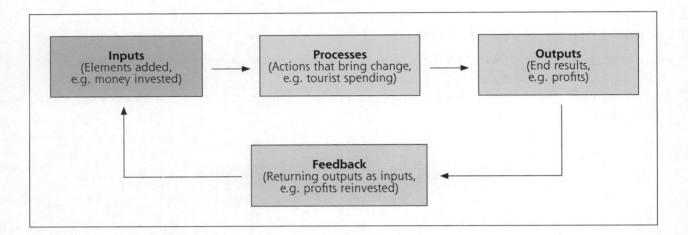

The operation of a cultural process has impacts on people and the environment. Economic processes provide positive impacts like incomes and jobs, which also have social impacts on health and wellbeing.

Any modification to the environment or increased use of an environmental resource by people could lead to negative impacts and a less attractive environment.

Since cultural processes **operate as a system**, feedback has an important role to play. For example, positive feedback in the form of reinvesting profits can lead to further development of new inputs like attractions and accommodation. Too many tourists can produce negative impacts like overcrowding and pollution of the environment, and lead to declining interest, reduced profits and less future investment.

Cultural processes constantly **change over time**, which creates **temporal variations**. There are changes in the elements and the process itself leading to changes in the environment. It is important to know how and why these changes have occurred. Changes in the types of tourists who visit a location would change the numbers and types of attractions that would be successful, which would also change the supporting facilities and infrastructure. For example, wealthier tourists generally prefer hotel accommodation with extra comfort and quality. They are more likely to spend a lot of money at tourist attractions, but also at retail shops and restaurants. In this way the cultural process constantly develops as new interactions occur and bring rejuvenation and renewal.

The **operation** of a cultural process also creates **spatial variations**, which can be seen in the spatial patterns in the elements across the environment. The location of the elements like attractions and accommodation types would depend on factors that would be the most beneficial for that business. For example, attractions might be located close to each other so tourists can visit multiple sites easily. Hotels tend to be located in city centres so tourists have better access to them as well as facilities like retail shops and restaurants. Spatial variations can be mapped and analysed to better understand the interactions in the cultural process. Clusters of activity can be identified in central locations for ease of accessibility or along major roads for the same reason.

     ISBN: 9780170446914

Tasks

1 Below is a systems diagram. Annotate the diagram using the examples in the list below.

Word list: Attractions, Tourist spending, Profits reinvested, Accommodation, Tourists, New attractions, New accommodation, Pollution, Traffic, Marketing, Facilities and infrastructure, Development of new attractions.

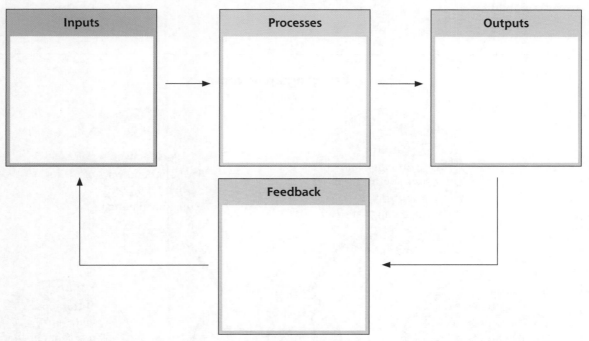

2 Explain how feedback from tourist perceptions could affect tourism in the future.

_____

_____

_____

_____

_____

_____

_____

3 Explain how feedback from high traffic congestion and pollution could affect tourism in the future.

_____

_____

_____

_____

_____

_____

_____

Key geographic concepts

Key geographic concepts are important because they will help build your understanding of the underlying themes or ideas in geography. When you have a conceptual understanding of a topic and examples, it becomes possible to transfer that understanding to new contexts.

The key geographic concepts are:

- Change
- Interaction
- Perspectives
- Sustainability.

- Environments
- Pattern
- Processes

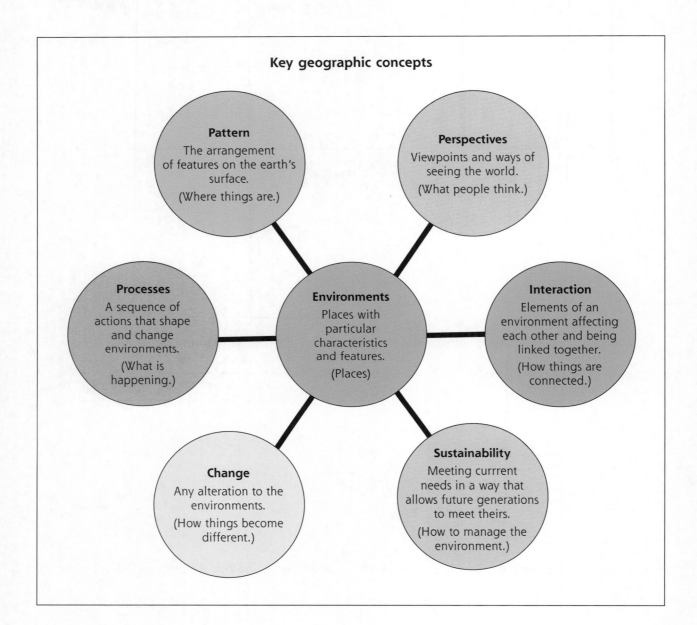

Key geographic concepts

Pattern
The arrangement of features on the earth's surface.
(Where things are.)

Perspectives
Viewpoints and ways of seeing the world.
(What people think.)

Processes
A sequence of actions that shape and change environments.
(What is happening.)

Environments
Places with particular characteristics and features.
(Places)

Interaction
Elements of an environment affecting each other and being linked together.
(How things are connected.)

Change
Any alteration to the environments.
(How things become different.)

Sustainability
Meeting currrent needs in a way that allows future generations to meet theirs.
(How to manage the environment.)

Key geographic concept — Change

Change is essential in geography because it applies to every process and environment and the effects can be seen in the contexts studied. It is any alteration to the environment, either spatial or temporal.

Task

Explain how the key geographic concept of change relates to the connections between the elements in the systems diagram on page 10.

_____

_____

_____

_____

_____

_____

_____

_____

_____

_____

_____

3

The tourism development process

Tourism development

In this chapter, we will examine the tourism development process. To begin, it is important to know what the key elements are and how they interact. This will help build an understanding of how the cultural process operates.

It is essential to note that tourism itself is not a process, it is an economic activity, and **tourism development** is the cultural process that we are studying.

Tourism development is about the interactions of elements due to the temporary visits of people travelling to destinations outside of where they normally live or work, primarily for recreation or leisure as tourists.

Tourism development is a process because:

- It describes how tourism changes over time. This involves new facilities being developed or changes in tourist types and numbers.
- It is made up of related activities which have interacting elements that shape the environment.
- The interactions are driven by people. This makes it a cultural process.
- It involves the exchange of money for goods and services. This makes it an economic process.

This is different to tourism, which was defined as people travelling away from their homes to places of interest for leisure or recreation. Tourism does not take into account the impacts that shape and change the environment.

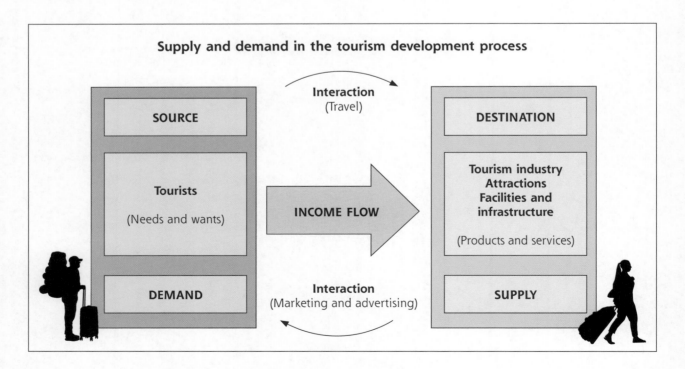

Supply and demand in the tourism development process

SOURCE	Interaction (Travel)	DESTINATION
Tourists (Needs and wants)	INCOME FLOW	Tourism industry Attractions Facilities and infrastructure (Products and services)
DEMAND	Interaction (Marketing and advertising)	SUPPLY

 ISBN: 9780170446914

Supply and demand are essential ideas in the tourism development process. Demand from tourist markets leads to the development of a local tourism industry to supply those needs, creating the interactions between elements. For example, demand for high-quality hotel accommodation could lead to more development of those types of facilities. If a location could not supply enough to meet the demand, it could lead to fewer tourists visiting and a loss of potential customers.

The diagram indicates that tourists will travel to a destination if it can meet their needs and wants. It is up to the destination location to supply the products and services to attract and retain the tourists. A key interaction in this process comes from marketing and advertising. This allows a destination location to reach out to potential customers and convince them to visit and spend money in the area.

When tourists spend money at an attraction, that business spends money on wages for staff. They might also need to source food, drink and other supplies from the local supermarket or stores. In addition, staff members spend their wages locally. All these interactions contribute to the circular flow of money through the local economy, which benefits the entire community.

The tourism development model

The tourism development model shows the key interactions between the source location, which is where the tourists come from, and the destination location, where the tourism development process occurs. There are also interactions between the elements, as their activities can complement each other like attractions and retail, for example, a glacier walk attraction and retail stores selling outdoor clothing.

At the destination, there are the key elements which drive the tourism development process. The attractions, tourism industry, facilities and infrastructure, and regulators are all linked and interact to support the conditions at a location that tourists would like to visit.

We will now examine the elements separately, using local examples as a case study. It is important to keep in mind the interactions between these elements at the same time.

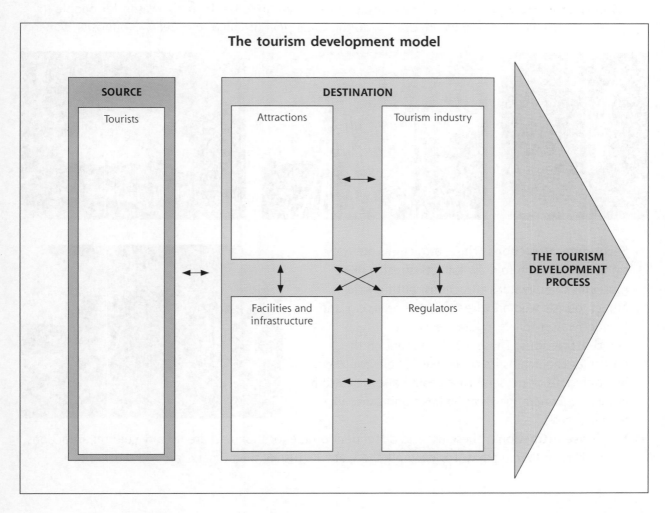

The tourism development model

SOURCE — Tourists

DESTINATION — Attractions, Tourism industry, Facilities and infrastructure, Regulators

THE TOURISM DEVELOPMENT PROCESS

📍 Task

Annotate the diagram on the previous page using the examples in the list below.

Word list: Beaches, Transport companies, Local council, Hotels, International tourists, Travel companies, Information services, Central government, Restaurants, Motels, Museums, Roads and highways, Domestic tourists.

Attractions

Attractions are the most important element in the tourism development process because they are the reason why tourists will travel to the destination location. Without tourists, there would be no tourism development. Attractions are places that have unique or interesting characteristics, which people want to visit.

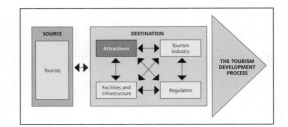

Attraction types

Attractions can be categorised into the following types:

- **Natural attractions**. These are natural features like beaches, glaciers, mountains, lakes, and forests.
- **Cultural attractions**. These are features or events made by people like museums, music festivals, theme parks, and historic monuments.
- **Primary attractions**. These attractions are significant enough to be the main reason for people to travel to a location, e.g. the Pyramids of Giza, the Great Wall of China, the Treaty grounds at Waitangi.

- **Secondary attractions**. These attractions are not significant enough to draw tourists directly, like museums or adventure attractions, but they do attract tourists who are already in the area, e.g. the Hundertwasser toilets in Kawakawa.
- **Fixed attractions**. These are dependent on the particular characteristics of the natural environment, like waterfalls or geysers. These could also be cultural attractions, where the location has significance like Stonehenge.

- **Footloose attractions**. These are less dependent on the location and are usually cultural attractions like casinos or theme parks. For example there are 12 Disney theme parks around the world.

Task

Complete the table below by identifying all the relevant attraction types that apply to each example. Give reasons for your choices.

Attraction types: Natural, Cultural, Primary, Secondary, Fixed, Footloose.

Attraction	Attraction type(s)	Reasons
Pohutu Geyser, Te Puia		
Mitai Maori Village		
The Blue Baths		
The Luge		

Attraction	Attraction type(s)	Reasons
Zorb		
Lake Rotorua		
Government Gardens		

Tourists

Tourists are individuals or groups of people who travel for the purpose of recreation or leisure. Tourists choose to spend time making the journey to a destination location and then spend their money when they are there.

Tourists buy goods and services when they travel from their homes in the source locations to the tourism destinations. This interaction between source and destination locations is central to the tourism development process, as new money from tourists enters the local economy providing incomes, jobs, and revenue for local government. Tourist spending also leads to investment in new attractions and the facilities and infrastructure that support them. This is how the tourism development process shapes and changes the environment.

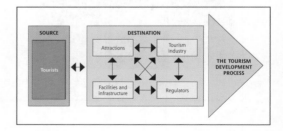

The main reasons tourists travel are for:

- Holidays and leisure
- Business activities
- Visiting friends and relatives.

Tourists can also be categorised as either international tourists who travel from outside the destination country or domestic tourists who travel from within the same country. There are important implications from this distinction, as international tourists bring money into the national economy. There might also be cultural and linguistic barriers to bridge as well as physical distance. Domestic tourists are also valuable because there are more opportunities for return visits without the same barriers.

Tourist motivators

There are key factors that influence tourist motivations to travel. They are:

- **Accessibility**. Ease of access to a destination by air or road links or by cruise ship reduces the time and cost of travel.
- **Affluence**. Tourists will need sufficient amounts of money to fund their activities and the free time to do them.
- **Awareness**. Knowledge of attractions and supporting facilities and infrastructure can influence decision-making. Marketing and advertising is used to reach new potential tourists.

These three motivators are important here because tourist awareness of, and accessibility to, attractions creates a demand and market for tourism-related products.

The reasons for travel and other factors such as age, affluence, home location, and personality type have impacts on elements in the tourism development process, as this affects the attraction types visited, accommodation needs, length of stay, and, ultimately, tourist spending.

Travel personality — Plog's continuum

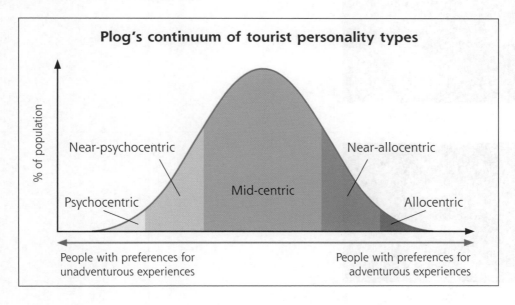

Plog's continuum shows a range of tourist personalities for a range of activities based on their underlying psychological preferences. The scale runs from psychocentric types, who prefer unadventurous experiences like planned group tours, to allocentric types, who are more interested in adventurous experiences. They are more likely to be free independent travellers (FITs) and go backpacking. The majority of people lie in the middle between the two opposites. Since this can vary due to a range of factors like age or culture, it is important that tourist destinations have a range of attractions and facilities to cater to different tourist needs and wants.

Table showing tourist preferences by adventurousness type

Tourist preferences	Psychocentric	Mid-centric	Allocentric
Attractions	Guided tours, gardens	Museums	Adventure tourism activities
Accommodation types	Hotels	Hotels, motels	Motels, backpacker hostels, campgrounds
Activities	Shopping	Eating out	Spontaneous travel
Planning	Pre-planned trips	Plan to have some new experiences	Less pre-planned trips

Task

Complete the table below by identifying all the tourist personality types that are most likely to visit each example. Give reasons for your choices.

Attraction	Tourist personality type	Reasons
Pohutu Geyser, Te Puia		
Mitai Maori Village		
The Blue Baths		

Attraction	Tourist personality type	Reasons
The Luge		
Zorb		
Lake Rotorua		
Government Gardens		

Facilities and infrastructure

Facilities are places that offer services and products which support the needs of tourists, like accommodation, retail, and hospitality.

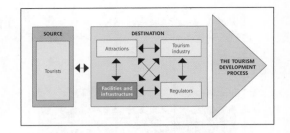

Infrastructure is about the systems and structures that support social and economic activity, like transport and communication networks, water and power. Infrastructure also supports the local economy, not just tourism activities.

Tourism development relies on more than just attractions. Without the facilities and infrastructure to support all tourist needs and wants, the tourism industry could not function. Although there are costs associated with infrastructure development, taxpayers and ratepayers in tourist destinations will benefit from tourist spending in their local areas as well as from improved infrastructure.

Infrastructure enhancements provided by public finances could include public transport, public toilets, free Wi-Fi in central locations, wider and better roads to accommodate large buses and coaches. Environmental restoration and enhancement projects to beautify an area also benefit local residents.

With the right conditions for businesses, new facilities can develop, and these can include new tourist attractions. This is how the tourism development process creates temporal and spatial variations in the environment. The next most important need tourists will have once they are in the area is accommodation.

Accommodation

There are different types of accommodation to suit the varying needs of both individuals and tourist groups. The needs of psychocentric tour groups will be very different to free independent travellers (FITs) looking for a backpacker's hostel or campground. These are differentiated by the services they offer, by location, and by price. By providing a number and range of accommodation types, a tourist destination can benefit from greater numbers of tourists by meeting the needs of diverse groups of tourist markets.

Here are the key details on accommodation types:

- **Hotels**. These tend to be larger, multilevel facilities with a range of room sizes. They have restaurants and often have bars, gyms, swimming pools, and function rooms for groups like international tourist groups and conference groups.

- **Motels**. These tend to be smaller scale without the extra facilities of hotels. Rooms have kitchens so visitors can do their own cooking. Motels need to have enough parking to accommodate their guests, who usually drive their own vehicles and tend to be domestic tourists.

- **Resorts/lodges**. These tend to be exclusive facilities with a small number of rooms and are hidden away in locations of natural beauty. They are defined by high-quality, luxury standards to attract wealthy tourists from all over the world.

- **Backpacker hostels**. These tend to be more basic in the services offered. Rooms are smaller and guests use communal facilities like kitchens, entertainment rooms, and bathrooms. These facilities suit the needs of FITs.

- **Campgrounds**. These are open spaces set up for tents and camping vans.

Task

Complete the table below by ranking the accommodation types in order by price from highest (1) to lowest (4). Give reasons for your choices. Hint: Think about the tourist types using them.

Ranking by price	Accommodation type	Reasons
	Hotels	
	Resorts/lodges	
	Backpacker hostels	
	Motels	

The tourism industry

The tourism industry includes any activity that provides goods and services and tourists. This can also include operators of attractions, accommodation providers and other businesses that cater to tourist needs like:

- Travel agents and websites
- Tourism information services
- Transportation providers
- International travel package providers.

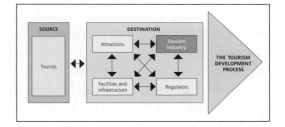

The local tourism industry also has connections to the global tourism industry through providers of online services like travel and booking websites such as Tripadvisor and Expedia. These websites allow tourists to plan their entire journey without the need for a travel agency. Tourists can also customise all the components of their journeys from accommodation and transportation to attraction visits and know how much these will cost.

A well-functioning tourism industry creates economic benefits for the local community from tourist spending. For example, while souvenir shops cater directly to tourists, bars and restaurants are likely to attract both tourists and local residents as customers.

Regulators

The regulators of the tourism development process are groups which make decisions about how tourism operates in a particular area.

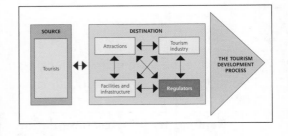

They are usually central or local government agencies or overseeing industry organisations. Nationally, tourism-related activities fall under the jurisdiction of the Ministry of Business, Innovation & Employment (MBIE), while local councils manage smaller-scale activities. Regulators are able to make and enforce rules that apply to groups participating in the tourism industry. Regulations might relate to health and safety, land-use zoning for activities, and environmental management like water-quality issues to ensure sustainable outcomes.

Regulators are generally not involved in operating tourism businesses, which typically remain private enterprises, but their actions influence activities, and these can promote the tourism development process and lead to continued economic development.

The government is the ultimate regulator and can shut down operations. For example, in 2020, international and at times domestic travel were shut down due to Covid-19, and Whakaari/White Island tourism operations were shut down over safety concerns after a volcanic eruption in 2019.

The tourism development process — summary of interactions

There are interactions between the source and destination elements in the tourism development process as well as among the destination elements themselves. These interactions lead to changes over time and changes in the environment. The way in which the elements interact together determines what the destination looks like at a particular place and time.

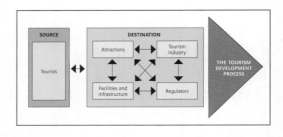

For example, as tourist numbers grow or change, it promotes changes to the attractions and facilities. As new attraction types develop, they may bring different types of tourists. As new facilities and infrastructures develop, they might encourage new types of business to open, which in turn may attract a different type of tourist. For example, a new attraction which draws wealthier tourists could lead to the development of additional hotels or resorts to meet their needs rather than backpacker hostels and campgrounds. Other services like bars and restaurants that would appeal to their tastes would also be in demand.

  ISBN: 9780170446914

Task

Explain how the key geographic concept of interaction relates to the connections between the elements in the tourism development model.

_____

_____

_____

_____

_____

_____

_____

_____

_____

_____

_____

Butler's model of tourism development

Butler's model of tourism development is a diagram which shows changes over time in the tourism development process.

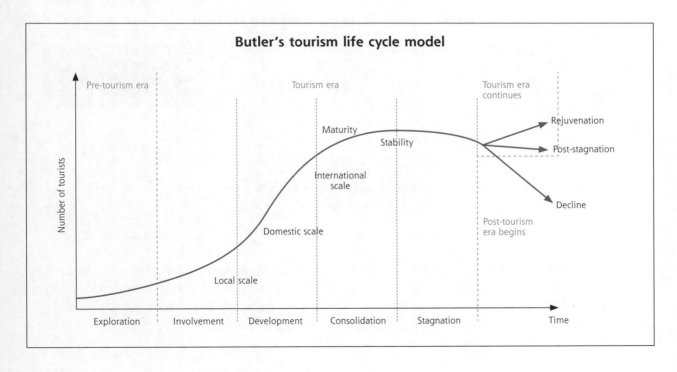

Butler's model allows us to visualise the stages in the life cycle of a tourist destination:

- **Exploration**. This involves the selection of a primary attraction around which a tourism industry and infrastructure can be based.
- **Involvement**. This involves the establishment of basic facilities and infrastructure like accommodation to support visitor numbers, even if it is only at a local scale.
- **Development**. This involves further establishment of facilities and infrastructure through increased investment. There is greater awareness of the destination from marketing and advertising, reaching a national scale.
- **Consolidation**. This involves the establishment of secondary attractions to make the most of the tourist numbers that visit the primary attraction.
- **Stagnation**. The tourist destination has reached maturity in terms of attractions and infrastructure and there is a lack of new growth or development.

Post-stagnation, there are two main possible outcomes:

- **Rejuvenation**. This involves changes in focus or in marketing to attract new tourists or to encourage previous visitors to return, allowing the tourist destination to develop again.
- **Decline**. This involves a decline in tourist numbers due to apathy or negative perceptions from experiences. This can lead to a negative feedback loop with closures and job losses.

The tourism development process is ongoing and the tourism industry needs to be responsive to the needs and wants of tourists. The interactions between the elements lead to change and temporal variations.

Cumulative causation

Cumulative causation is the process which creates the changes seen in Butler's model. It acts like feedback in a system, which is how it has been modelled below. If a location becomes a popular tourist destination, it will develop new elements over time, which is cumulative causation.

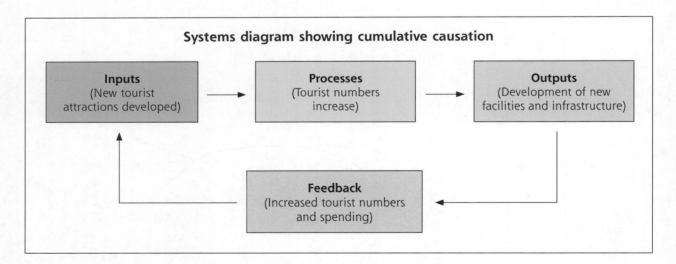

Attractions are seen as inputs in a system. They draw tourists who travel to the area for new experiences and they spend money in the area. If they have a positive experience (feedback), they are more likely to return or tell others, generating more income. This income can be reinvested to develop new and existing attractions, but also facilities and infrastructure to support the tourism industry.

Agglomeration

Tourism development is an ongoing economic process where the tourism industry has developed to supply elements and features in the form of attractions and facilities and infrastructure in order to meet the demands of visiting tourists. This is known as agglomeration, which is a concentration of similar activities in the same area. This leads to spatial variations seen in the environment.

This is an ongoing process and the tourism industry needs to be responsive to tourist needs in order to ensure an ongoing income stream. Positive feedback loops allow the industry to continue to invest in the destination, leading to cumulative causation as the tourism industry develops over time.

Circumstances from outside the process can also effect change in the form of government regulations of activities or travel restrictions. An economic downturn like the global financial crisis of 2008 or especially the global Covid-19 pandemic of 2020 substantially reduced tourist numbers and spending at destinations.

Rotorua city's attractions are complemented by the many services that tourists need like accommodation, restaurants and retail.

Tasks

1 Write definitions of the following terms in your own words.

Tourism development: _____

_____

_____

_____

Supply and demand: _____

_____

_____

_____

Cumulative causation:_____

_____

_____

_____

Agglomeration:_____

_____

_____

_____

Key geographic concept — Processes

Processes are a sequence of actions that shape and change environments. They are either natural or cultural and create the spatial and temporal patterns seen. If the sequences of actions do not occur in the expected order, this could affect the operation of the process and the effects on the environment.

2   Explain how tourism development operates as a process.

_____

_____

_____

_____

_____

_____

_____

_____

_____

_____

_____

Tourism development in Rotorua

4

In this chapter, we will examine all the elements in tourism development by looking at different examples throughout the Rotorua District. This will help to build our knowledge of the case study area and our understanding of the interactions between the elements.

Rotorua location

Rotorua is a city in the Bay of Plenty region. It is centred on the southern shore of Lake Rotorua. The lake is situated in a caldera, which was formed in a major volcanic eruption approximately 240,000 years ago. The depression left behind filled up to become the lake, while Mokoia Island in the centre is a lava dome.

Rotorua's location in the Taupo Volcanic Zone has contributed to the overall landscape as well as the geothermal activity that started the tourism development process in the area. The rest of the chapter will examine the operation of the tourism development process in Rotorua.

The simplified map on page 30 shows the key landmarks in the city. The main two areas are:

- The core. This is the centre of activity and there is a concentration of facilities and infrastructure that supports the tourism industry and civic life.
- The periphery. This includes all the areas outside the core. There is less concentration of facilities and infrastructure, but there may be clusters of activity in some locations.

The Central Business District (CBD) was established in 1881 to be the core focus of settlement near the lakefront. Fenton Street is the main street, and this runs from the lake to Whakarewarewa and Te Puia, which is the centre of geothermal activity in the township and primary attraction.

Large hotels are located near the lakefront, while cheaper motel accommodation can be found along the length of Fenton Street.

State Highway 5 to the west is the main road link into Rotorua for tourists coming from Auckland, while tourists coming from Tauranga would take State Highway 36, which meets up with State Highway 5.

Secondary attractions which require large areas of land can be found on the periphery of the city, but along State Highway 5, giving them accessibility and visibility.

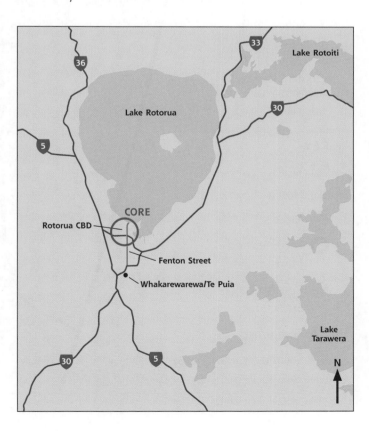

Map showing key features of tourism development in Rotorua

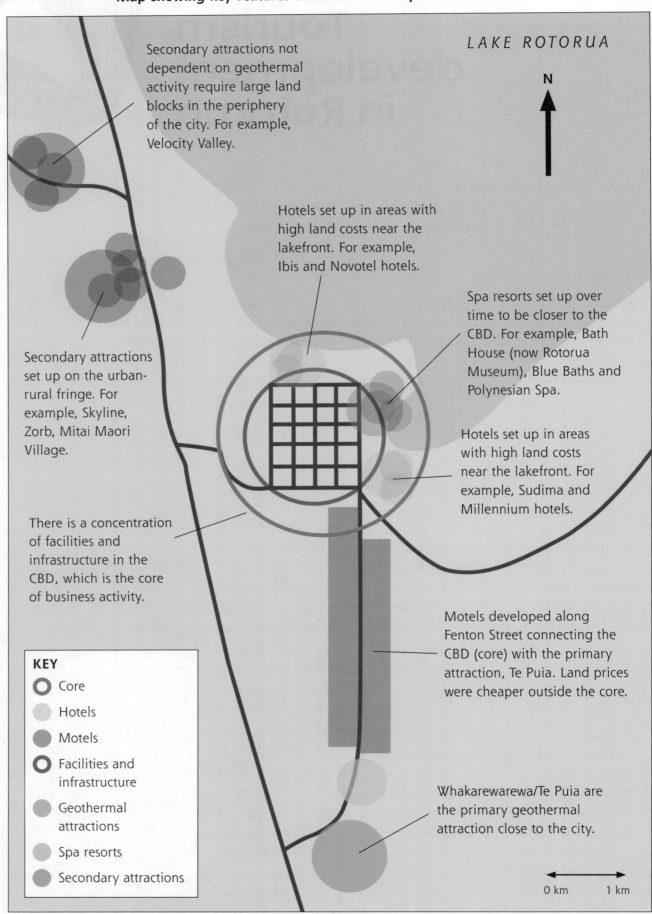

LAKE ROTORUA

N

Secondary attractions not dependent on geothermal activity require large land blocks in the periphery of the city. For example, Velocity Valley.

Hotels set up in areas with high land costs near the lakefront. For example, Ibis and Novotel hotels.

Spa resorts set up over time to be closer to the CBD. For example, Bath House (now Rotorua Museum), Blue Baths and Polynesian Spa.

Secondary attractions set up on the urban-rural fringe. For example, Skyline, Zorb, Mitai Maori Village.

Hotels set up in areas with high land costs near the lakefront. For example, Sudima and Millennium hotels.

There is a concentration of facilities and infrastructure in the CBD, which is the core of business activity.

Motels developed along Fenton Street connecting the CBD (core) with the primary attraction, Te Puia. Land prices were cheaper outside the core.

Whakarewarewa/Te Puia are the primary geothermal attraction close to the city.

KEY
- ◎ Core
- ○ Hotels
- ● Motels
- ◎ Facilities and infrastructure
- ● Geothermal attractions
- ● Spa resorts
- ● Secondary attractions

0 km   1 km

  ISBN: 9780170446914

Rotorua elements

Below is a table with some information about the key features of tourism development in Rotorua.

Feature	Description
 Te Puia	Activity at Te Puia is mainly centred on the naturally occurring geothermal features in the area. It is home to Pohutu Geyser, which is the largest in New Zealand. There are also cultural attractions here based on Maori culture and this is best seen in the Maori Arts and Crafts Institute, which teaches traditional art forms in public view.
 Mitai Maori Village	Mitai Maori Village seeks to give tourists an experience of Maori culture, with guided walks through a replica of a traditional pa, followed by entertainment and a hangi dinner.
 Rotorua i-SITE Visitor Information Centre	This is a central location where tourists can find relevant information about services in Rotorua and make bookings for attractions and accommodation, etc. It is also the main stop for intercity bus services to Rotorua.
 Skyline	Skyline includes the gondola ride up Mt Ngongotaha to the luge tracks next to the main building, which has a cafe and restaurant. There is a cluster of other activities that tourists could do, like the Skyswing and zip-lining.
 Motels	There are many mid-price options for accommodation, with motel accommodation set up along the length of Fenton Street from the CBD down to Whakarewarewa. There are also hotels at various points here too.

Feature	Description
 Rotorua Lakes Council	The Rotorua Lakes Council is the local authority that is in charge of local infrastructure like roading and development. They set the rules for land-use zoning.
 Energy Events Centre	This large events centre is used for exhibitions, concerts and conventions and is located in the CBD. It also hosts sporting fixtures like the ANZ Championship netball games.

 # Task

Based on your own knowledge and research of Rotorua, add the names of specific examples to the tourism development diagram below, e.g. i-SITE Visitor Information Centre in the tourism industry.

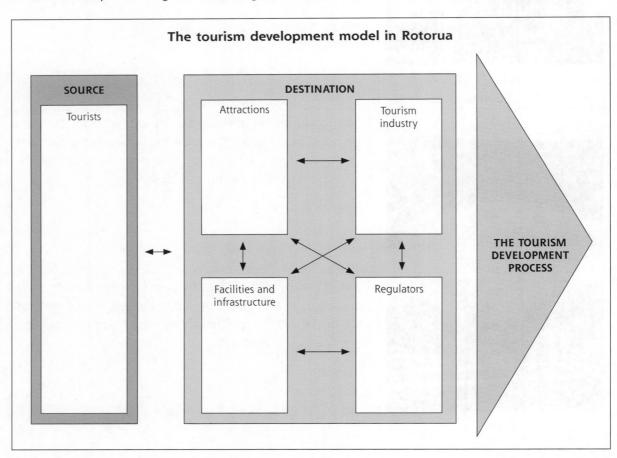

The tourism development model in Rotorua

SOURCE

Tourists

DESTINATION

Attractions

Tourism industry

Facilities and infrastructure

Regulators

THE TOURISM DEVELOPMENT PROCESS

Temporal variations

5

Tourism development is an ongoing process, and temporal variations refer to changes over time. Changes occur in all the elements, from tourists who visit and the attractions they want to see, to the facilities and infrastructure that support the tourism industry. All of which creates new interactions in the tourism development process. It is important to understand the reasons why there is change.

The changes in the Rotorua tourism industry can be represented using Butler's model (page 25). It models the changes over time in the tourism development process from early development to maturity and market saturation. Rejuvenation or decline are possible outcomes. Cumulative causation is also in effect, as a tourism destination develops over time if there is continued tourist interest and investment in new attractions and facilities and infrastructure.

Phases of development

There are three main phases of development, which roughly correspond to the stages of Butler's model.

- Phase 1 runs from 1830s to the 1930s, and can be further divided into the Pioneer phase (1830–1886) and the Consolidation phase (1886–1930s).
- Phase 2 runs from the 1950s to 1980.
- Phase 3 runs from 1980 to the present day, and can be further divided into the Diversification phase (1980–1987) and the Specialisation phase (1987 to the present day).

Tourism development phases in Rotorua

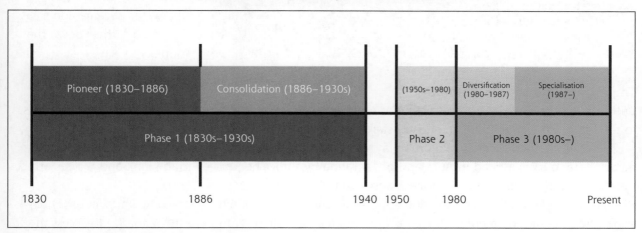

We will now examine all the phases of development and the key changes that create temporal variations.

Phase 1: The development phase (1830s–1930s)

The Pioneer phase (1830–1886)

The Rotorua area had long been settled by Te Arawa iwi, among others, with nature providing abundant resources from the forests and lakes. The geothermal waters of the region provided accessible hot water, which could be used for cooking.

European settlers and missionaries arrived in the area in the 1830s, with the first known tourist, naturalist John Bidwell, visiting in 1839.

The Pink and White Terraces were the primary attraction for the area and were often called the Eighth Wonder of the World. These were large silica deposits that had built up as terraces over time as silica-laden water flowed downhill from one pool to another, creating the distinctive terraces. The colours reflected trace minerals in the water.

The Pink and White Terraces were a unique natural landscape and the remoteness of the location in New Zealand made them a must-see attraction, which would draw wealthy international tourists with the time and money to make the very long journey.

Tourists would bathe in the pools formed by the terraces, and they could find a temperature that suited their preferences as the water became cooler as it moved further down the slope.

The tourism industry in Rotorua

With the Pink and White Terraces acting as a primary attraction and a strong factor in drawing tourists to the region, the local tourism industry started to develop.

There was limited access to Rotorua via overland routes by coach once the usually wealthy northern hemisphere tourists arrived in Auckland or Tauranga by ship. The journey to the Pink and White Terraces went across both land via horse coaches and water via canoe.

As can be seen on the map on page 35, travellers to the region initially stayed at Ohinemutu (1), an established Maori settlement on the lakefront, and then travelled through to Te Wairoa (2) to access the terraces across Lake Rotomahana (3). Accommodation developed in Te Wairoa, and there was a need for tour guides and transportation to the terraces.

 ISBN: 9780170446914

Map showing the locations of early tourism in Rotorua

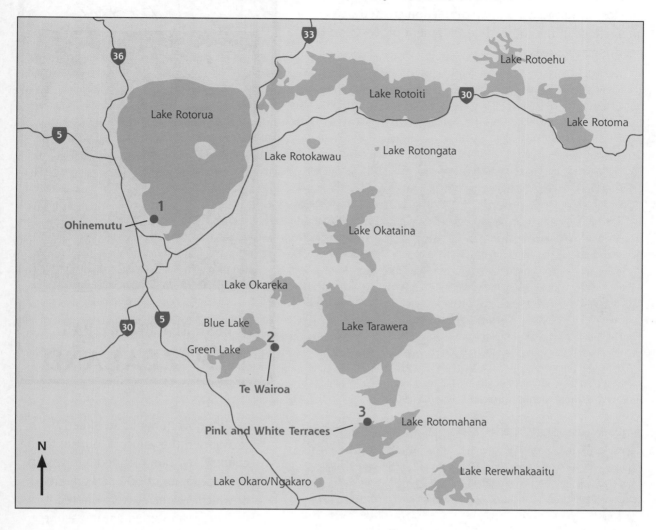

Tourism was exclusively an activity of the wealthy, so there was little domestic tourism at the time. Tourists had to be adventurous and would fit under the allocentric end of tourist personality types.

The Mt Tarawera eruption, 1886

On 10 June 1886, there was a huge eruption at Mt Tarawera and the Pink and White Terraces were destroyed. The eruption also buried Te Wairoa village in ash and killed up to 150 people in the surrounding areas.

The site of the Pink and White Terraces was believed to have been found submerged in Lake Rotomahana in 2011 by GNS scientists.

The Consolidation phase (1886–1930s)

The loss of Rotorua's primary attraction created new interactions in the tourism development process and the need for new ideas.

Visitors to Tikitere thermal attraction, Rotorua 1908.

The government had already recognised the value of tourism to the country and had planned a township of Rotorua about 600 acres in area, of which 125 acres were offered for European settlement under the Thermal Springs District Act of 1881.

This would allow for the development of facilities and infrastructure as well as new attractions. The area was completed in 1894 where the CBD is today, and this helped increase tourist numbers in the town following the destruction of the Pink and White Terraces. A railway from Auckland to Rotorua was completed in the same year, which also contributed to greater tourist numbers in the town due to improved accessibility.

In addition to physical infrastructure, the government, as a regulator, gave the Department of Tourism and Health Resorts control of the town in 1907 to oversee the new spa resorts the government was relying on as new attractions.

New attractions

There were still other geothermal attractions in the area that could be developed. Te Waimangu Geyser became active after the Mt Tarawera eruption and was the world's largest geyser at the time. Eruptions continued until 1908 when the geyser was declared extinct.

The Waimangu Rift Valley was created by the Mt Tarawera eruption and it continues to be a natural attraction in the present day. Tourists can even take a boat cruise on Lake Rotomahana over the locations where the Pink and White Terraces are thought to be submerged.

Whakarewarewa became the primary geothermal attraction, with Pohutu Geyser at its heart. There has been a permanent Maori settlement in this former fortress (Te Puia) for centuries.

The biggest advantage Whakarewarewa has is its location close to Lake Rotorua and the newly established township. This gives greater accessibility for tourists.

Health and spa resorts

At the centre of the government's tourism strategy was the Bath House, completed in 1908. Thermal waters were pumped into the building for bathing and massage, as the mineral waters were believed to cure chronic ailments.

The hope was for Rotorua to entice wealthy tourists from the northern hemisphere to the 'Great South Seas Spa'. The Tudor-style wood building was situated in the beautiful Government Gardens, close to the township, and epitomised the ideas of health and wellbeing.

After the Second World War, health spas waned in popularity, the building proved to be expensive to maintain, and the once state-of-the-art Bath House was ultimately a failure, having never become a profitable business. After restoration work, it reopened as the Rotorua Museum in 1969. The museum closed in 2016 due to the need for improved earthquake strengthening measures to meet current standards of safety. In such an old large building, this has proved expensive and also shows the role of regulators and the effects.

Other spa resorts

Spa baths were popular in the 1930s, with two other sites opening up, and in close proximity to the Bath House near the CBD. The Ward Baths (now the Polynesian Spa) opened in 1931. The Blue Baths, which opened in 1933, was the first bath house in New Zealand to offer mixed bathing.

The First World War, the Great Depression, and the Second World War saw dramatic drop-offs in tourist numbers globally. Rotorua was stagnating and needed to change.

Rotorua Bath House just after it opened. Over the years the Bath House has required extensive maintenance work, including earthquake strengthening.

Tasks

1 Discuss how the loss of the Pink and White Terraces brought about change in Rotorua's tourism industry.

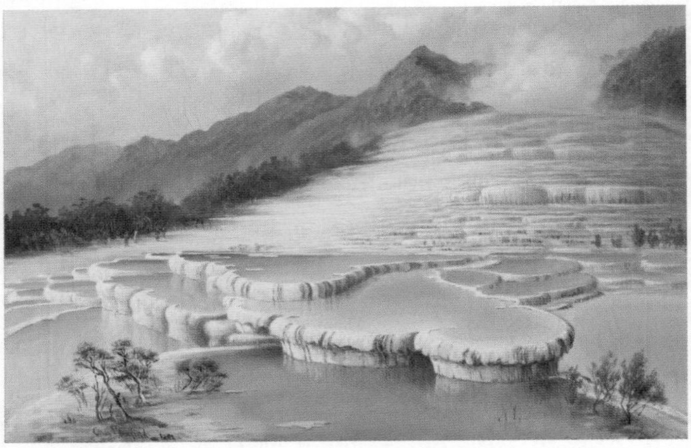

_____

_____

_____

_____

_____

_____

_____

_____

_____

_____

_____

_____

_____

_____

_____

2 Describe the role of central government in establishing tourism development in Rotorua and explain their motivations.

_____

_____

_____

_____

_____

_____

_____

_____

_____

_____

3 On the blank map below, locate and label the features that were developed in Phase 1 of tourism development in Rotorua. Annotate the map with the reasons why those features were established. Give the map a title.

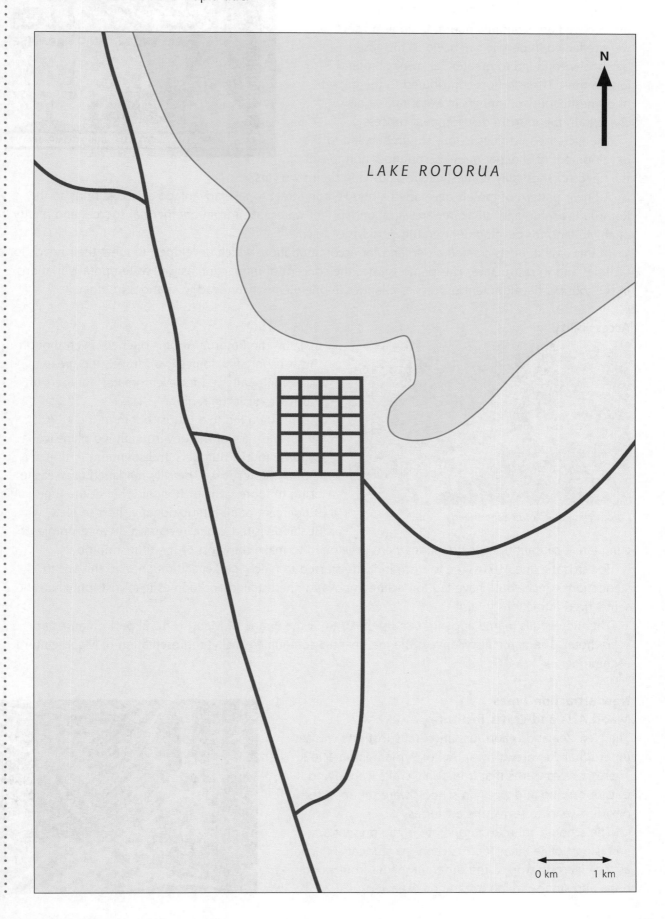

Phase 2: Mass tourism (1950s–1980)

Post Second World War affluence

After the Second World War, New Zealand experienced an economic boom. Changes in society were reflected in people's behaviour. Increased personal wealth led to greater car ownership and leisure time. These factors contributed to the growth of domestic tourist numbers in Rotorua and new changes in the tourism development process.

The increase in domestic tourist numbers led to the development of cheaper motel accommodation to suit the needs of short-stay domestic tourists driving their own cars.

A linear pattern of motels developed along Fenton Street to take advantage of accessibility to the Rotorua township and Whakarewarewa, abundant hot water taken from geothermal sources and plenty of cheap land for development on the outskirts of town.

In this case, tourists created a demand for accommodation, which developed to meet their needs for facilities and infrastructure. This demonstrates the concept of interaction as well as temporal variations in the tourism development process as changes in one element can lead to changes in others.

Accessibility

Construction of Rotorua airport, early 1960s.

In 1964 the Rotorua Airport opened. Even though this was only for domestic air travel, the greater accessibility allowed for even greater numbers of tourists to visit the region.

This then led to a tourist boom in the 1970s, which was part of the government-led plans to increase tourist numbers and spending in the country. Rotorua was heavily marketed overseas to bring in more tourists. It is said that at the time, 90 percent of all international visitors to New Zealand visited Whakarewarewa. In an example of cumulative causation, new attractions were developed to make the most of the tourist numbers.

Towards the late 1970s, the tourism industry started to reach market saturation, and this led to stagnation, which could have led to decline. Innovation had stalled and domestic tourists only had the same, tired attractions to visit.

The 'Rotovegas' name appeared and seemed to fit the image of Maori cultural performances at every hotel. The name also conveys the inauthentic, sometimes 'tacky' representation of Maori culture at the time.

New attraction types
Maori Arts and Crafts Institute

The New Zealand Maori Arts and Crafts Institute was set up at Whakarewarewa by an Act of Parliament in 1963. There were concerns that traditional crafts were dying out, so carving and weaving schools were set up in the next few years to revive the art forms.

The school continues to operate in the present day and is part of Te Puia. Tourists can view students in action as they learn to work using different media from wood, bone and stone carving to raranga (weaving).

 ISBN: 9780170446914

Rainbow Springs

Fairy Springs, which opened in the 1930s, used the local spring water as a home to introduce trout. This was integrated with neighbouring attraction Rainbow Springs in 1967.

The famous nocturnal kiwi house was soon added as well as other buildings and facilities. Operations expanded in 1985 with the addition of Rainbow Farm, which added an agriculture element, until this was sold off and Kiwi Encounter was opened. Rainbow Springs was closed for a time during the Covid-19 pandemic because New Zealand's borders were closed to international tourists.

The changes over time at Rainbow Springs provide further evidence of cumulative causation in the tourism development process as the tourism industry adapted to changing conditions and attempted to meet the needs of different tourist markets.

Agrodome

The Agrodome, which opened in 1971, is another example of a changing tourism industry in Rotorua and was a rare example of a secondary attraction that wasn't geothermal or a Maori cultural attraction. Set in a 140-hectare working cattle and sheep farm, it is famous for its farm show, which demonstrates shearing and milking and features sheep and dogs.

The Agrodome receives approximately 350,000 visitors per year, mostly international tourists, and there are translation services for tour groups that cater to the language needs of visitors.

The Agrodome is an iconic tourist attraction, but it fell victim to changing market conditions and was closed for a time during the Covid-19 pandemic while New Zealand's borders were closed to international visitors. Both Rainbow Springs and Agrodome are owned by Ngai Tahu Tourism. This iwi organisation is based in the South Island but they have tourism ventures throughout New Zealand and demonstrates the diversity of groups in the tourism industry in Rotorua.

Tasks

1 Describe the differences between Phase 1 and Phase 2 of tourism development in Rotorua.

_____

_____

_____

_____

_____

2 Explain the role of cumulative causation in the tourism development process in Phase 2.

_____

_____

_____

_____

_____

3 On the blank map below, locate and label the features that were developed in Phase 2 of tourism development in Rotorua. Annotate the map with the reasons why those features were established. Give the map a title.

N

LAKE ROTORUA

0 km   1 km

OPERATION OF TOURISM DEVELOPMENT IN A NEW ZEALAND ENVIRONMENT   ISBN: 9780170446914

Phase 3: Diversification and specialisation (1980 to the present day)

Tourism industry changes

By the 1980s, the tourism industry in Rotorua was beginning to stagnate, as seen in Butler's model.

In the late 1970s–1980s, many of the geothermal features, especially the geysers, seemed to wane in power and intensity. Research suggested that the high use of geothermal water through private bores in homes and businesses was draining off the groundwater that fed these features.

In addition to this, there was a sense that there was nothing new to see in Rotorua, especially for domestic tourists. It was a case of 'been there, done that'.

The government had directed and regulated tourism development in the region since the early days. However, by the mid-1980s, government ideology and priorities had shifted, with the government allowing the private sector to take a more active role in the tourism industry.

Attraction changes — adventure tourism

There was a move away from the usual geothermal and Maori cultural attractions to new 'adventure' tourism attractions. This was intended to rejuvenate the tourism industry in Rotorua by bringing in tourists from new markets as well as enticing domestic tourists to return.

Skyline opened in 1985 along State Highway 5, which is the main road into Rotorua for tourists travelling from Auckland. It is a distinctive feature that is highly visible to newly arrived tourists, and it is easily accessible.

The core attraction revolves around the gondola, which now transports 600,000+ visitors up the side of Mt Ngongotaha every year. At the top of the gondola is the 'gravity-powered' luge, which hosts over 1.8 million rides a year and has operated since 1987.

Skyline has continued to change by adding more attractions and upgrading the facilities. In 2005, the gondola capacity was increased, and a cluster of secondary attractions continues to be added. The full list of auxiliary products includes: Skyswing, zip-line, mountain biking, star gazing, the Jelly Belly store, iNZpired store, volcanic hills, market kitchen, Stratosfare Restaurant.

Skyline now has operations in Queenstown and overseas in Singapore, South Korea and Canada. This shows the ability of companies to expand their businesses into the international tourism industry especially in the case of footloose attractions.

ISBN: 9780170446914   PHOTOCOPYING OF THIS PAGE IS RESTRICTED UNDER LAW.

Cultural tourism

A desire for more authentic cultural experiences provided opportunities for Maori entrepreneurs in the tourism industry. Hotels had long run Maori cultural shows with a meal, but there was a sense of cultural appropriation, hence the 'Rotovegas' nickname.

Tamaki Tours, which opened in 1995, was unique in its day by setting up an authentic-looking pre-European Maori village on the outside of the city. The main drawcards are the cultural show and the hangi dinner, which follows tikanga Maori and manaakitanga, the Maori concept of hospitality and caring for visitors.

They run their own bus services to pick up and drop off visitors, which can add to the experience while also removing distance and accessibility as a barrier for tourists.

Events

Large-scale events offer opportunities for Rotorua to attract tourists. This will have flow-on effects throughout the local economy. This was shown by the Lions tour in 2017 when the Lions played the Maori All Blacks at Rotorua International Stadium. It was estimated this game brought in $11.3 million of visitor expenditure on accommodation, food and drink and visits to attractions while fans of both teams were in the city.

The Covid-19 pandemic of 2020 saw a sudden end to many large-scale events across New Zealand. Even without full lockdown restrictions, limits on crowd sizes and physical distancing requirements made such events uneconomical. This had flow-on effects with lower visitor numbers and spending. The Rotorua Marathon around Lake Rotorua is the oldest marathon event in Australasia, and in 2020 it did take place with thousands of runners and supporters involved.

Mudtopia Festival

The Mudtopia Festival was held at the end of 2017 at Arawa Park Racecourse and based on an event held in South Korea. The concept involved promoting a fun event with music and food and drink while showcasing Rotorua's credentials as a spa and wellness destination.

The event ran into controversy for its cost to ratepayers as it ran at a loss and for importing mud for the event from South Korea. Concerns were about biosecurity and a lack of promotion of local mud and products.

Crankworx

Mountain biking has become a big drawcard for Rotorua with tracks at Skyline and Whakarewarewa (Mountain Bike Rotorua 2011). This is part of the industry strategy of diversification to attract new tourist markets, making use of the natural environment.

Crankworx is the premier event in mountain biking, with four events held across the world. The Rotorua event is held on the slopes of Mt Ngongotaha, in partnership with Skyline, bringing in large numbers of visitors. Crankworx have signed a 10-year contract to keep the event in Rotorua.

Eco-tourism

The tourism industry in Rotorua has continued to adapt to market needs and conditions, sometimes responding to change and at others, initiating change.

In 1987, the government spent $21 million to clean up Lake Rotorua, which was previously used as a dumping ground for raw sewage and effluent run-off from surrounding farms. These clean-up efforts continue, with a $72.1 million government contribution to ongoing conservation efforts in 2008.

The change in focus can also be seen in new attractions, which focus on niche markets. For example, Canopy Tours, which opened in 2012, guides tourists around on zip-lines while promoting conservation.

Another example of the interaction of tourism and conservation is Wingspan. This is the national bird of prey centre and was established in 2002. It is a footloose secondary attraction where tourists can see native birds of prey like the karearea or New Zealand falcon up close as well as in the air during a daily show. This attraction educates the public about birds of prey and raises funds for ongoing conservation efforts.

With greater consumer consciousness about carbon emissions and environmental standards, it is important that tourists do not perceive travel to a location to be damaging to the environment. This adds an extra dimension to meeting tourist needs and wants.

The Rotorua Sustainability Charter is a group of tourism operators who have agreed to collaborate as a network to promote sustainable practice within the local tourism industry. Membership requires tourism operators to undergo an annual review on their sustainability and commit to new initiatives for future improvements.

Tourist changes

Greater personal mobility and greater accessibility allowed for backpackers to travel relatively cheaply. These free independent travellers (FITs) have their own interests and needs.

They tend to fit the more allocentric tourist personality type, are younger, and are more likely to stay at backpacker types of accommodation to save money.

Kiwi Paka, now known as the Backyard Inn, is typical of the cheaper backpacker-style accommodation and has been in business since 1986. The facilities are basic with communal bathrooms and kitchens. This is designed to cater to budget travellers, and this can be seen in the price of accommodation.

Asian tourist markets also became more important to the industry from the 1990s as economic development brought more disposable income and a desire to travel. Rotorua had to adapt to suit the needs of new tourist markets like Japan, China, and South Korea.

Cruise ship tourists are a new market for Rotorua despite the distance from the sea. Tour groups disembarking in Tauranga have the option to make day trips inland to Rotorua. Apart from the benefits

of there being more tourists, these groups tend to visit during the weekdays, which increases demand in off-peak times. Cruise ship tourists tend to prefer visiting natural attractions and there is always a possibility that they will return for a full trip in the future.

The Covid-19 pandemic had a severe impact on the international cruise tourism industry. Negative perceptions of cruise ships as unsafe increased following reports of hundreds of passenger infections on cruise ships. The pandemic blocked access to international tourists who would typically visit Rotorua. With the New Zealand border closed to non-citizens, domestic tourists became crucial in supporting Rotorua's tourism industry. This was accompanied by a change in marketing strategies to attract domestic tourists, and discounts were offered to further encourage tourist spending.

Regulator changes

Although the government had left the tourism industry as an owner/operator, it still has a role as a regulator with interests in social, economic and environmental impacts.

In 1986, the government decided to close all the bores within a 1.5-kilometre radius of Pohutu Geyser. Private homes and hotels/motels drawing the free geothermal waters were believed to be affecting the geyser's activity by reducing the available water pressure.

The Rotorua District Council followed up on the changes and the private bores were finally closed in 1987, although some businesses (hotels/motels) were exempt.

This is typical of how the government continues to interact with the tourism development process in Rotorua. There has since been an increase in geothermal activity around the geysers, with Pohutu Geyser having more frequent and long-lasting eruptions. In 2000/2001, the geyser erupted continuously for over 250 days.

Facilities and infrastructure changes

All of the changes have been part of the rejuvenation of the tourism industry in Rotorua as seen in Butler's model.

In the 1990s, the Rotorua District Council spent $35 million on revitalising the CBD including the i-SITE Visitor Information Centre. These upgrades also benefited local residents. The Rotorua i-SITE information centre was refurbished in 1993. This is funded by local government and provides free information.

Change does not always mean new additions. New conditions can mean old ideas become economically unviable. This can be seen with the cancellation of passenger rail services to Rotorua in 2001. The former railyards in the CBD made way for the Rotorua Central Mall which opened in 2000, and this lies at the heart of retail activities in the city.

Other changes in transport include short-lived international flights to Sydney from 2011 to 2015. Although they started with much fanfare, the service was mostly used by locals enjoying greater accessibility and choice and did not deliver on the expected increase in tourist numbers from the Australian market.

The Rotorua Energy Events Centre opened in 2007 and has 10 spaces to host sports events and conferences, adding to Rotorua's capacity. It is also another home venue for the Waikato Bay of Plenty Magic netball team, which plays in the ANZ Championship.

In 2013, 'Eat Streat' on Tutanekai Street was developed to be the hub of dining in Rotorua. The covered way and below-floor geothermal heating is a unique experience and it is situated in the CBD, but also close to the lakefront.

Accommodation changes

Airbnb was established in the US in 2008 and has spread around the world. It is a platform which connects hosts with tourist guests and offers a huge range of accommodation options, many at very economical prices.

At the top end of the price scale are the luxury resorts dotted around the region. These offer seclusion and privacy due to the costs, capacity and isolated surroundings.

For example, Treetops Lodge is surrounded by native forest, while Solitaire Lodge overlooks Lake Tarawera. These types of accommodation have been established to cater to the small niche market of wealthier tourists who are willing to spend extra money on luxury products and services, adding to the diversity of tourists and markets.

Tasks

1   Describe the differences between Phase 2 and Phase 3 of tourism development in Rotorua.

2   Explain the role of agglomeration in the tourism development process in Phase 3.

3   Explain how external factors like global pandemics can bring change to the tourism development process.

4 On the blank map below, locate and label the features that were developed in Phase 3 of tourism development in Rotorua. Annotate the map with the reasons why those features were established. Give the map a title.

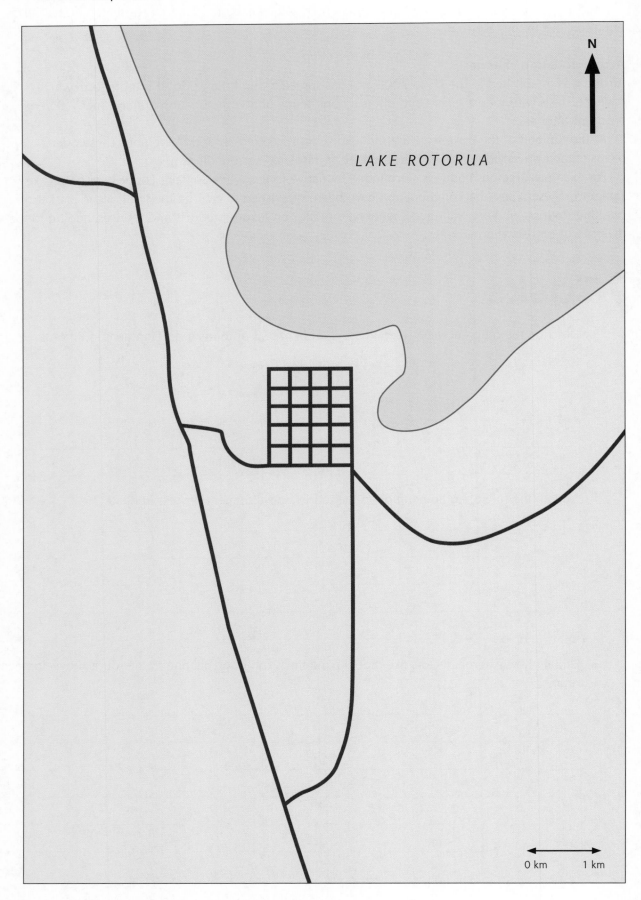

5 Complete the timeline by adding the key events and development of key attractions and facilities to it from the list below.

Timeline of key events in the tourism development process in Rotorua (1839–2020)

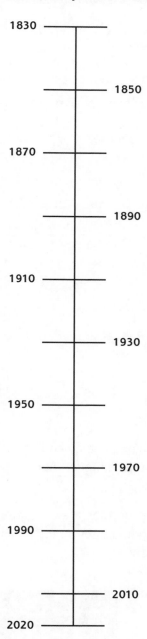

1830

1850

1870

1890

1910

1930

1950

1970

1990

2010

2020

Rotorua timeline of key developments

- First tourist to Rotorua (1839)
- Rotorua township established (1881)
- Pink and White Terraces destroyed (1886)
- Railway to Auckland built (1894)
- The Department of Tourism and Health Resorts takes control of Rotorua (1907)
- Government Gardens, Bath House (1908)
- Ward Baths, now Polynesian Spa (1931)
- Maori Arts and Crafts Institute (1963)
- Rotorua Airport (1964)
- Rainbow Springs (1967)
- Rotorua Museum (1969)
- Agrodome (1971)
- Skyline sky rides (1985)
- Kiwi Paka (1986)

- Bores closed, Luge (1987)
- i-SITE Visitor Information Centre (1993)
- Zorb (1994)
- Tamaki Tours (1995)
- Agroventures, now Velocity Valley (1998)
- Rotorua Central Mall (2000)
- Rotorua Energy Events Centre (2007)
- Mitai Maori Village (2008)
- First cruise ship tourist visits (2011)
- Canopy Tours (2012)
- 'Eat Streat' (2013)
- Redwoods Rotorua (2015)
- Crankworx (2016)
- Covid-19 pandemic (2020)

⦿ Tasks

1   Explain how the key geographic concept of perspectives relates to the changes seen in the tourism development process.

_____

_____

_____

_____

_____

_____

_____

_____

_____

_____

_____

_____

2   Annotate the diagram on the next page by adding key events listed below and explaining how they fit the relevant stage of Butler's model. Add a title to your diagram.

Event list:
- The Pink and White Terraces become a tourist attraction.
- Transport and accommodation are set up for the Pink and White Terraces.
- Rotorua township is established.
- Mass tourism begins.
- High government involvement and regulation in the tourism industry.
- Less government involvement and deregulation in the tourism industry.
- Development of adventure tourism.
- New tourist markets are encouraged.
- Diversification and specialisation.

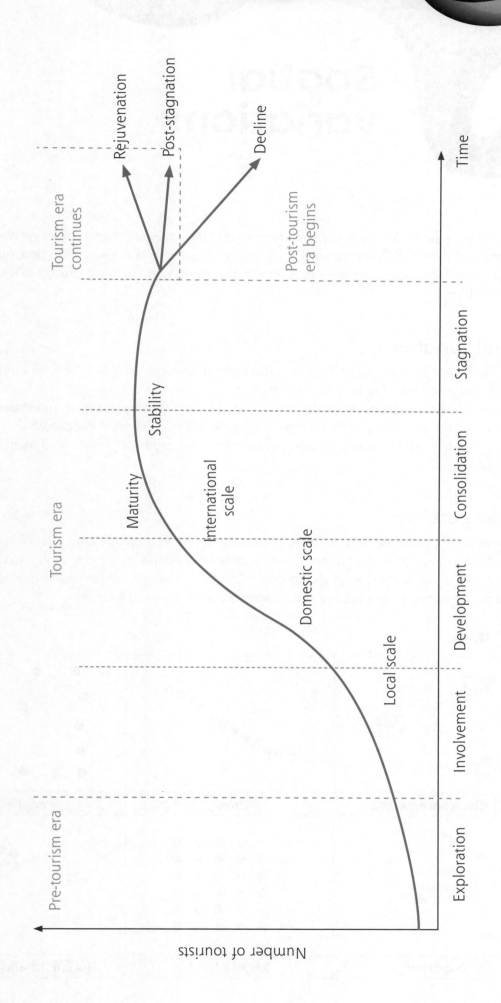

6

Spatial variations

The operation of the tourism development process over time has led to spatial patterns in the environment. There are different patterns depending on the elements in the tourism development model.

It is important to understand the interactions in the tourism development process that create the spatial variations.

Spatial variations

Tourism development as a process operates differently in different parts of the city and region, creating the spatial patterns seen. This is spatial variation.

By analysing how and why the patterns occur, we will have a comprehensive understanding of how the elements interact which lead to the operation of the tourism development process.

We need to link our knowledge of the spatial patterns created to our understanding of how the process works.

Spatial patterns

The operation of the tourism development process can create new spatial patterns and modify existing ones in the environment as the tourism industry continually adapts to the ever-changing demands of tourists. For example, there is agglomeration of secondary attractions along State Highway 5 to take advantage of the greater accessibility and awareness provided by the location.

Pattern types

Cluster/concentrated	Linear	Peripheral
Dispersed	Regular/grid	Nucleated and isolated

Task

Examine the maps on pages 54, 55 and 58 and use them to complete the table below.

Spatial patterns in Rotorua

Elements/ features	Pattern type	Examples	Reason for the pattern
Geothermal attractions			
Hotels			
Motels			
Backpackers/ hostels			
Secondary attractions			

Map showing the location of hotels in Rotorua

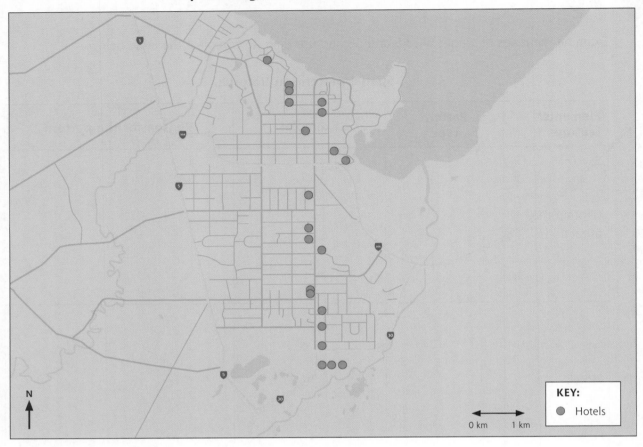

KEY:
● Hotels

0 km   1 km

N

Map showing the location of motels and backpackers/hostels in Rotorua

KEY:
● Motels
● Backpackers/
  hostels

0 km   1 km

N

Map showing the location of geothermal attractions in Rotorua

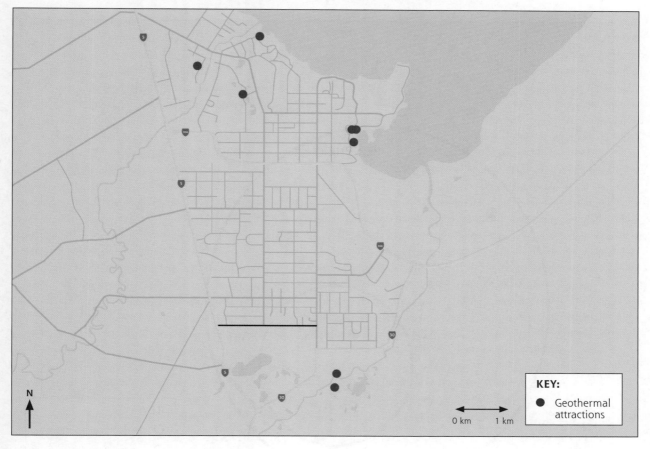

Reasons for spatial variations

In order to comprehensively analyse spatial variations, it is necessary to link our knowledge of the patterns to the elements and key geographic concepts of the tourism development process.

The core–periphery model

The core–periphery model emphasises the importance of the **core**, which is at the centre of development as seen in the map on page 56, prior to 1886.

The **periphery** is located away from the core and is on the edge of development.

Phase 1 spatial patterns

The core acts as a central hub for tourism-related activities in the environment. These are usually found around a central point, which acts as a gateway or entry point to the area.

Tourism development in Rotorua started with the geothermal natural attractions. These are dispersed throughout the environment due to the natural factors that created them.

ISBN: 9780170446914   PHOTOCOPYING OF THIS PAGE IS RESTRICTED UNDER LAW.

Map showing the core-periphery model of tourism development in Rotorua before 1886

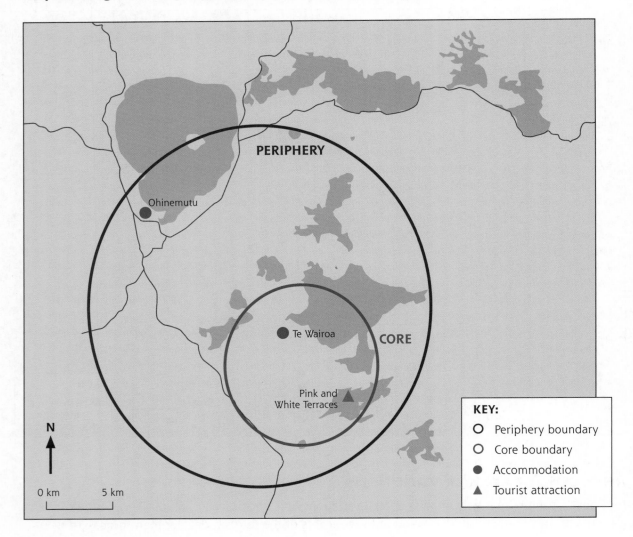

With the loss of the Pink and White Terraces, focus shifted to Te Puia as the primary geothermal attraction. This was combined with the spa resort developments closer to the township like the Bath House (1908).

The concentration of facilities and infrastructure in Rotorua was already under way when the Pink and White Terraces were destroyed. The government had purchased land to set up the township in the current CBD location in 1881. This was part of their strategy to develop the tourism industry in the region.

Accessibility is a key factor in the development of the core, as transport links like road (State Highway 5), rail (1894) and airports (1964) allow for the development of tourist attractions and the facilities and infrastructure that support it.

Over time, the size and scope of the core and the services offered expands, which is part of cumulative causation. Private businesses also moved to the CBD in the form of hotels, which provided accommodation for wealthy tourists who came to Rotorua to visit the health spas.

This clearly demonstrates supply and demand as the tourism industry develops over time and creates the spatial patterns seen in Rotorua.

Phase 2 spatial patterns

Greater affluence in the domestic tourist market after the Second World War led to an expansion of tourism activities. Tourism-related activities began to expand outside the core into the urban-rural fringe.

The linear pattern of motels and hotels that developed on Fenton Street stretching from the southern end of the CBD towards Whakarewarewa is a consequence of the tourism development process. This was due to greater availability of space needed for motels and cheaper land prices outside the CBD.

The linear pattern of accommodation also provided improved accessibility to the facilities and infrastructure in the CBD and the geothermal attractions centred on Whakarewarewa. There are also clear signs of agglomeration due to the concentration of facilities and infrastructure in Rotorua.

Phase 3 spatial patterns

The diversification of tourist attractions created new spatial patterns in the periphery. The cluster patterns of secondary attractions show clear signs of agglomeration as attractions are located together, such as the cluster around the Agrodome (1971), and Skyline (1985).

This ties in with temporal variations according to Butler's model as the tourism industry went through a period of rejuvenation.

The other advantage to these locations is ease of accessibility along State Highway 5, which is the main transport route into town, and large plots of cheap available land on the urban-rural fringe.

These attractions are highly visible, thus generating more tourist demand and business.

With rejuvenation, there has been an increase in the number of tourism-related activities being set up, as well as greater diversification and specialisation. This has led to spatial variation even within the same elements of the tourism development process.

This can be seen in the spatial patterns of accommodation. With higher land prices in the core, centred on the CBD and lakefront, only hotels can afford to establish themselves here, as seen in the maps. Hotels can build upwards to increase capacity and make the most of their capital investment.

Even though backpackers and hostels are at the other end of the price spectrum, the same logic applies in terms of building up. They would need a smaller land footprint, so can be located in the CBD.

It is important for hostels/backpackers to be located near key transport links which meet the needs of the tourist markets. In this case, FITs who would use buses to travel.

This can be seen in the example of the Rotorua Downtown Backpackers, which is located on Fenton Street, right next to the visitor centre and main bus stop. This also means that key attractions and amenities are within walking distance.

Other operators such as the Backyard Inn are located further away where land is cheaper, like near Kuirau Park or outside the city.

Resorts are different again, with these being located in a dispersed pattern across the region to take advantage of secluded prime locations.

The lack of accessibility is an advantage for the tourist markets who stay at resorts, as exclusivity is desirable and allows the businesses to charge more for the experience.

This clearly demonstrates spatial variation as the tourism development process operates differently across the environment. This has led to the various spatial patterns seen across the Rotorua environment, as all supply-side elements in the tourism industry adapt in order to meet the demands of the different tourist markets to encourage greater visitor expenditure.

This core is centred on the CBD in the present day, while the periphery extends beyond the city to encompass the geothermal attractions and exclusive resorts in the region, further than what is seen in the map below.

The clusters of secondary attractions located near Skyline and the Velocity Valley are on the urban-rural fringe, which is the boundary between urban and rural areas.

Map showing the core-periphery model of tourism development in Rotorua in the present day

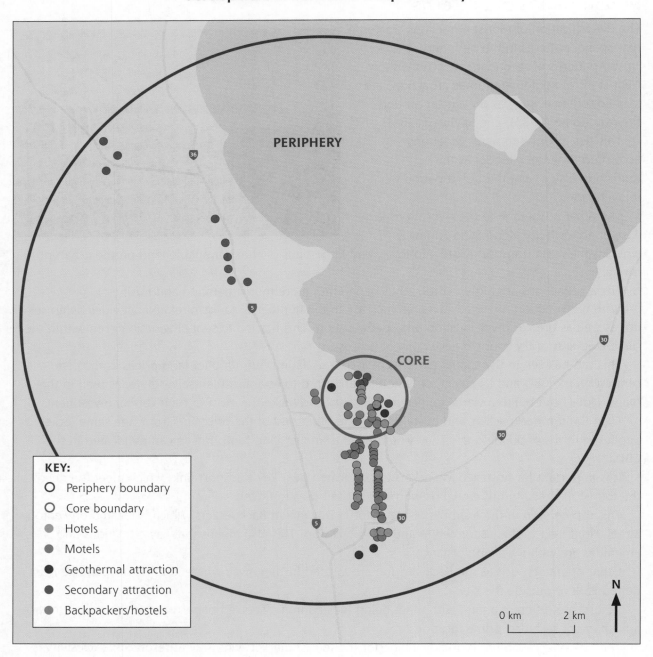

KEY:
- ○ Periphery boundary
- ○ Core boundary
- ● Hotels
- ● Motels
- ● Geothermal attraction
- ● Secondary attraction
- ● Backpackers/hostels

0 km   2 km

N

       ISBN: 9780170446914

⦿ Tasks

1 Describe the different spatial patterns that have developed in the tourism development process in Rotorua.

_____

_____

_____

_____

_____

_____

_____

_____

2 Explain why the tourism development process operates differently across Rotorua, creating the spatial patterns seen.

_____

_____

_____

_____

_____

_____

_____

_____

_____

Key geographic concept — Pattern
Patterns are the arrangement of features in the environment in the case of spatial patterns.
Temporal patterns are the changes in characteristics over time.

3 Explain how the key geographic concept of pattern relates to features of tourism development seen in the Rotorua environment.

_____

_____

_____

_____

_____

_____

_____

4 The table below organises the spatial patterns seen by time period so we can also see the temporal variations.

Complete the table by explaining the links between the spatial patterns and the operation of the tourism development process in Rotorua.

Time period	Pattern type	Link to tourism development	Explanation of link	Interacting elements
Phase 1	**Dispersed** (Geothermal attractions)	**Butler's model — development** **Awareness**		
	Concentrated (Facilities and infrastructure)	**Cumulative causation** **Supply and demand**		
Phase 2	**Linear** (Hotels/motels)	**Accessibility** **Affluence**		
Phase 3	**Concentrated** (Backpackers/ hostels)	**Accessibility** **Supply and demand**		
	Clusters (Secondary attractions)	**Butler's model — rejuvenation** **Agglomeration**		

5 On the blank map below, locate and label the spatial patterns seen in the tourism development process in Rotorua. Annotate the map with the reasons why those spatial patterns developed. Give the map a title.

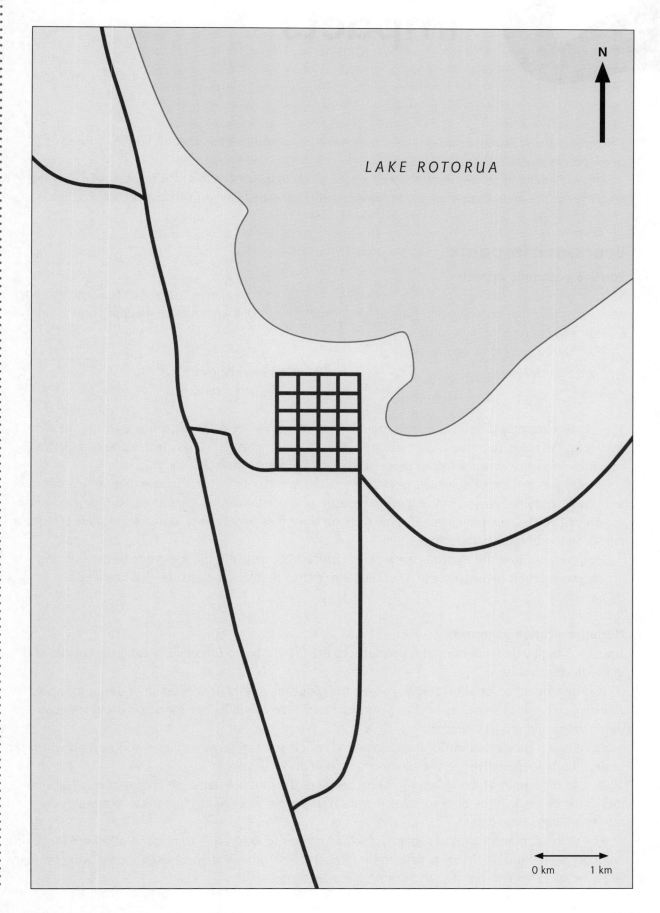

LAKE ROTORUA

N

0 km   1 km

7

Tourism impacts

The temporal and spatial variations seen in Rotorua are a result of the operation of the tourism development process.

This wide-ranging process impacts on all aspects of life in the city. In this final chapter, we will examine both the positive and negative effects on the economic, social and environmental aspects of Rotorua.

Economic impacts

Positive economic impacts

Tourism generates hundreds of millions of dollars in total visitor expenditure in Rotorua. The revenues are categorised into domestic and international expenditure. This is all money that has come from the outside and been spent in Rotorua.

The main spending categories are:

- Food and beverage
- Retail
- Accommodation
- Recreational services.

The tourism industry developed in Rotorua in order to supply the demands of the tourists. The more successful the industry is, the more tourism expenditure increases. This has created the cumulative causation seen as tourism activities cluster themselves around Rotorua.

Twenty-five percent of Rotorua's workforce is either directly or indirectly (jobs that service the tourism industry) employed in the tourism industry and is estimated to contribute 10.5 percent of the region's GDP. The economic benefits flow through the entire city, even in industries that are not directly part of the local tourism industry.

However, because the tourism development process is constantly changing due to the ongoing action and reaction between supply and demand, tourist number and income flows cannot be guaranteed.

Negative economic impacts

Tourism is highly dependent on the perspectives the tourists have. Therefore, it can be a volatile and unpredictable industry.

It is dependent on political stability across the globe, as this influences people's desire to travel. Tourists can be price sensitive, so changes in exchange rate can influence spending decisions about destinations and services needed.

As part of a globalised world, the tourism industry is susceptible to negative world events which can impact tourist perspectives on the safety or affordability of travel.

Due to the nature of the tourism industry, many of the jobs are part-time and seasonal. This means there may be work in the busy summer months, but people will need to find other jobs when the tourist numbers slow down.

The tourism industry uses this to attract school groups to Rotorua. This helps to offset the low tourist numbers even if group discounts have to be offered. These groups are organised for the winter months, which helps to smooth out demand during the off-season.

 ISBN: 9780170446914

At times, the tourism industry can be a victim of its own success. The large influx of tourist numbers requires continued spending on infrastructure from both central and local government to keep pace with increased needs.

There is a continual need for marketing and branding and advertising to international tourists. All this comes from taxpayer and ratepayer money and it can be difficult to quantify the effectiveness and value of these activities.

The global Covid-19 pandemic of 2020–2021 highlighted the vulnerability of the tourism industry to external shocks. Even without border closures and travel restrictions, tourist demand would also be lower due to job losses and insecurity in source countries. Without international tourists, Rotorua relied more on domestic tourists, especially from Auckland. However, the increased domestic demand was not enough to offset the lost spending from international tourists.

Social impacts

Positive social impacts

While attempting to generate revenue from visiting tourists, the development of infrastructure, services and the hosting of events can benefit local residents as well. The development of 'Eat Streat' on Tutanekai Street in the CBD in 2013 provides an enjoyable dining experience for locals.

The existing spa resorts are great amenities for health and wellbeing for local residents. New plans for the Wai Ariki Hot Springs and Spa near the lakefront will add to the area's reputation as the spa and wellness capital of the southern hemisphere.

The locals also benefit from the hosting of major events in the city. Often there are discounts or special access for local residents. Key examples are Crankworx since 2016 and Mudtopia in 2017. Others include Rotorua Marathon, Rotorua Blues Festival, Super Rugby matches, the ANZ Netball Championship games, and concerts.

Cultural impacts

Fifty-seven percent of tourists associate Maori culture with tourism in Rotorua. This has presented opportunities for Maori entrepreneurs to develop authentic Maori cultural attractions such as Tamaki Tours in 1995 and Mitai Maori Village in 2008.

These are family businesses and employ Maori as tour guides and performers. This also educates tourists about Maori culture and values in a more direct way so that visitors have a greater appreciation of tikanga Maori and local traditions.

The Maori Arts and Crafts Institute, which opened in 1963 and is now an integral part of Te Puia, has helped with the revival and maintenance of many traditional art forms. These are on display for tourists to see and experience. Apprenticeships are provided for young Maori artists.

Negative social impacts

The large numbers of visitors can put pressure on infrastructure and services during peak season. Traffic congestion increases and it can be difficult to find accommodation. In addition to this, most attractions are operating at full capacity during the summer months, leading to increased waiting times.

As well as the positive social impacts for Maori culture, there are concerns that by commercialising aspects of culture, this can lead to the loss of meaning and mana. Aspects of culture become tokenistic for the sake of tourist audiences. Plastic tiki and other souvenirs can be seen as classic examples.

There are also negative perspectives that these cultural tourism attractions could also have the effect of commercialising Maori culture and reducing the essence of tikanga Maori to mere song and dance. Despite the clear interest in the indigenous culture of New Zealand, international tourists could leave with the perception that Maori culture is static while ignoring the changes due to the influences of colonisation and modernisation thanks to their interaction with these attractions.

Environmental impacts

Negative environmental impacts

With over 3 million tourists visiting Rotorua annually, there are bound to be negative impacts on the environment as a result of the tourism development process.

The large tourist numbers put pressure on local infrastructure leading to increased waste, rubbish and sewage disposal. There is increased air pollution from the traffic congestion, especially with buses and coaches.

Rotorua generates more waste for a city of its size due to visiting tourists. The infrastructure needed must still be paid for by local ratepayers and this can put pressure on services.

High visitor numbers can cause damage to the environment along walking tracks from high use or damage to vegetation if people do not stick to the tracks.

Having a township develop on top of a geothermal field has created its own issues. Local residents were taking free hot water from their own bores. This reduced the overall underground water pressure, which reduced the activity of Pohutu Geyser. The council-ordered closure of the bores in 1987 has seen geyser activity increase.

Positive environmental impacts

There is the need to preserve and maintain the environment in order to avoid negative impacts on tourism in Rotorua.

In 1987, the government spent $21 million to clean up Lake Rotorua, which was used as a dumping ground for raw sewage and effluent run-off from surrounding farms. These clean-up efforts continue with a $72.1 million government contribution to ongoing conservation efforts in 2008.

The lakefront has been transformed into an attraction and is the home for lake-based activities such as the Katoa Jet Boat to Mokoia Island, Rotorua Duck Tours, the *Lakeland Queen* boat cruises and Volcanic Air scenic flights.

Other eco-tourism attractions have been developed, such as Redwoods Tree Walk, Canopy Tours, and Wingspan, which is New Zealand's national bird of prey conservation centre.

These attractions leverage a greater consciousness about sustainability and the value of a pristine natural environment as a tourist attraction. This can help to boost visitor expenditure and create positive perspectives about Rotorua.

📍 Tasks

1 Do the positive impacts of tourism development outweigh the negatives? Justify your response.

_____

_____

_____

_____

_____

_____

_____

_____

Key geographic concept — Environments
Environments are places with particular characteristics and features. These are shaped by interactions and processes which create the patterns in the environments.

2 Explain how the key geographic concept of environments relates to the tourism development process.

_____

_____

_____

_____

_____

_____

_____

_____

Key geographic concept — Sustainability

Sustainability is about meeting current needs in a way that allows future generations to meet theirs. It is important to minimise any negative impacts on environments.

3　Explain how the key geographic concept of sustainability relates to the impacts of the tourism development process.

_____

_____

_____

_____

_____

_____

_____

_____

_____

_____

4　Complete the summary table of impacts below.

Effect	Economic	Social	Environmental
Positive			
Negative			

5 On the blank map below, locate and label key features of the tourism development process that have positive and negative impacts in Rotorua. Annotate the map with the reasons for these impacts. Give the map a title.

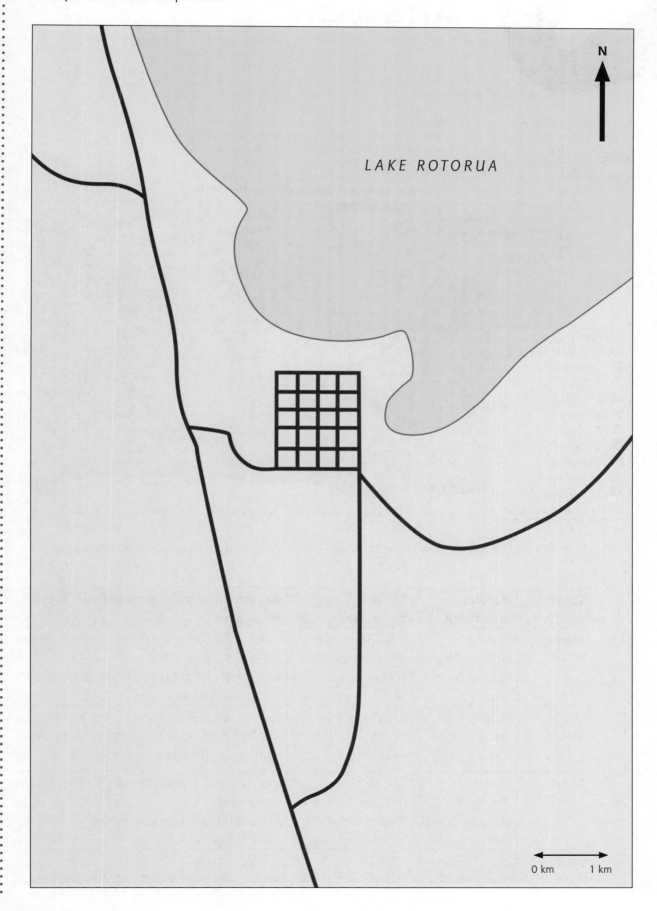

Answers

Chapter 1
(pp. 5–6)

1 a

Top 10 tourist destinations in the world

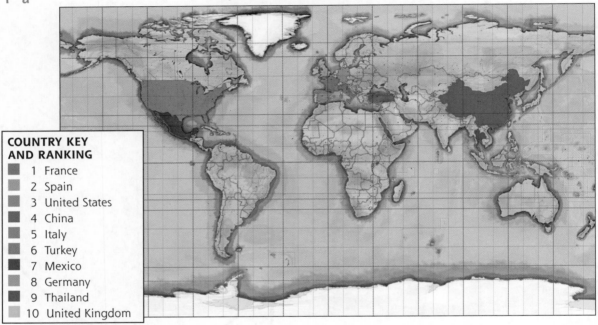

COUNTRY KEY AND RANKING
- 1 France
- 2 Spain
- 3 United States
- 4 China
- 5 Italy
- 6 Turkey
- 7 Mexico
- 8 Germany
- 9 Thailand
- 10 United Kingdom

b All the countries are located in the northern hemisphere and there are clusters in Europe.

2 a

Country	Natural attractions	Cultural attractions	Reasons why people would visit these attractions
France	The Alps, beaches on the Mediterranean Sea	Eiffel Tower, Arc de Triomphe, the Louvre	Winter and summer experiences with natural attraction. Important historic sites and art from European culture.
United States	Yellowstone National Park	Las Vegas, Disneyland, Statue of Liberty	Unique volcanic landscapes and vast forests. Modern attractions that cater to large numbers of travellers including families.
China	Himalayas, Zhangjiajie National Forest Park	The Forbidden City, the Great Wall of China, the Terracotta Army	Mountainous areas and unique cave and river landscapes. Ancient historic sites which have great cultural significance.

b There are many attractions to visit and these countries are accessible and encourage tourism.

1

Image	Process	Process type: Natural or Cultural
	Horticulture	Cultural
	Aeolian deposition	Natural
	Agriculture	Cultural
	Vegetation succession	Natural
	Forestry	Cultural
	Electricity generation	Cultural

2 Because the inputs have been heavily modified by people, for example animals for agriculture are selectively bred by farmers, and bananas require people to plant and grow them. Forestry plantations are set up by people to grow only one type of tree.

(p. 11)

1

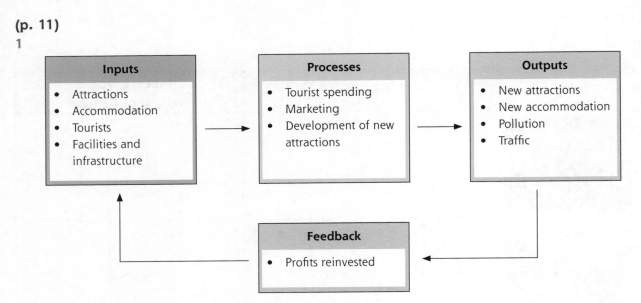

2 Positive tourist experiences could lead to return visits and increased word of mouth encouraging friends and relatives to visit in future. Negative tourist experiences could result in fewer return visits and negative reviews and word of mouth would potentially discourage new visitors.

3 Negative tourist experiences due to traffic congestion or a less attractive environment due to pollution could decrease visitor numbers in future.

(p. 13)

The key geographic concept of change relates to the connections between the elements in the systems diagram because the elements are linked together and affect each other. When there is a change in one element, there will be changes in others and change across the system. Change is essential to the tourism development process because of the perspectives and actions of people.

Chapter 3
(p. 16)

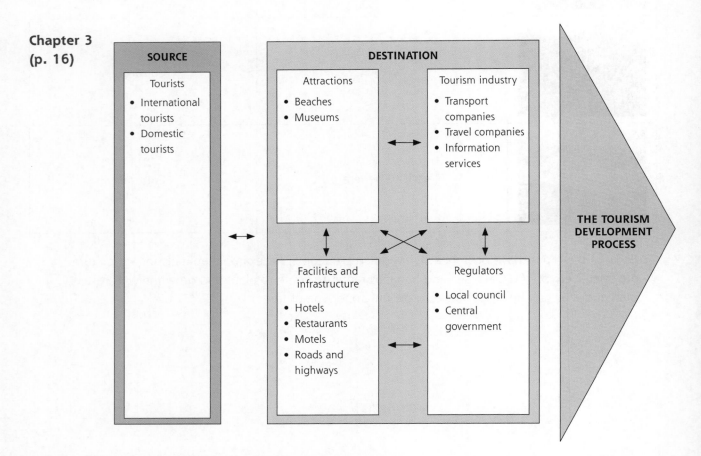

(pp. 17–18)

Attraction	Attraction type(s)	Reasons
Pohutu Geyser, Te Puia	Natural Primary Fixed	Geothermal activity depends on natural processes at fixed locations. Pohutu Geyser is significant because it is New Zealand's largest geyser.
Mitai Maori Village	Cultural Secondary Footloose	An attraction demonstrating Maori culture is about people. It is unlikely to attract tourists on its own. The attraction could be located somewhere else.
The Blue Baths	Cultural Secondary Fixed	It is cultural because people are using the geothermal waters. It is unlikely to attract tourists on its own. The attraction depends on the geothermal waters below.
The Luge	Cultural Primary Footloose	It is cultural because people are using the hillslope. It is a significant attraction for certain markets on its own. The operators have similar attractions in other locations.
Zorb	Cultural Secondary Footloose	It is cultural because people make the Zorbs out of plastic and roll them down a hill. It is unlikely to attract tourists on its own. This attraction could be set up anywhere with a hill.
Lake Rotorua	Natural Secondary Fixed	The lake is a naturally occurring feature even if the surrounding environment has been modified by people. It is unlikely to attract tourists on its own. This attraction's location is determined by the natural processes that made it.
Government Gardens	Cultural Secondary Footloose	The gardens were shaped and maintained by people using introduced plants. It is unlikely to attract tourists on its own. This attraction could be located anywhere and was chosen for its central location.

(pp. 20–21)

Attraction	Tourist personality type	Reasons
Pohutu Geyser, Te Puia	Mid-centric	Although there is a small element of danger from the geothermal activity, the landscape is quite safe and requires a short walk to get around the site.
Mitai Maori Village	Mid-centric	This attraction is focused on Maori cultural performance and a hangi meal. Some sense of adventure is needed to embrace cultural differences.
The Blue Baths	Mid-centric	Going to thermal pools is supposed to be a relaxing experience, but public bathing could be unsettling for some people.
The Luge	Allocentric	The luge is designed for high speed and thrills and there is a risk of crashing.
Zorb	Allocentric	The Zorb can be unsettling due to the rotation and closed space.
Lake Rotorua	Psychocentric	Some of the activities near the lake might be adventurous, but a walk around the usually calm lakefront is a relaxing option.
Government Gardens	Psychocentric	The gardens are a relaxing place to spend time. It is usually a quiet space with lots of places to take photos.

(p. 23)

Ranking by price	Accommodation type	Reasons
2	Hotels	Based on price and large group travel from international tourists.
1	Resorts/lodges	Based on price and seclusion.
4	Backpacker hostels	Based on price and basic needs of free and independent travellers.
3	Motels	Based on price and needs of domestic tourists.

(p. 25)

The key geographic concept of interaction relates to the tourism development process because the process does not occur without the connections between different elements in both source and destination locations. Interactions between the elements will guide the overall success of tourism activity in a destination location in meeting the needs and wants of different types of tourists.

(pp. 27–28)

1 **Tourism development:** This is about the interactions of tourism-related elements which are designed to meet the needs and wants of tourists.

   Supply and demand: This is about the connection between the goods and services people have and what they want. This relates to prices.

   Cumulative causation: This is about the build-up of activities in a location over time because the conditions are favourable to this.

   Agglomeration: This is about the location of similar activities near each other to take advantage of higher customer numbers.

2 Tourism development operates as a process because it requires a sequence of actions to occur in the tourism-related elements to happen which then leads to tourists travelling and spending money in a tourism destination. This then leads to social and economic benefits through ongoing tourism development.

Chapter 4

(p. 32)

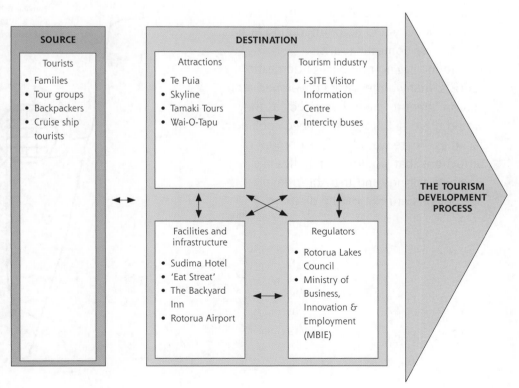

Chapter 5
(pp. 38–39)

1 Because the Pink and White Terraces were a primary attraction, alternative attractions were needed to bring tourists to the region, or this would lead to a decline in the tourism industry and local economy.

2 The central government was heavily involved in the establishment of the tourism development process because they had money to set up businesses and buy land. They also made the laws and regulations around tourism.

3 See answers at right on the diagram.

KEY
- O Core
- Hotels
- O Facilities and infrastructure
- Geothermal attractions
- Spa resorts

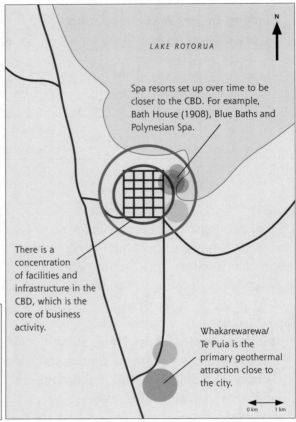

Map showing temporal variations in Phase 1 of tourism development in Rotorua

LAKE ROTORUA

Spa resorts set up over time to be closer to the CBD. For example, Bath House (1908), Blue Baths and Polynesian Spa.

There is a concentration of facilities and infrastructure in the CBD, which is the core of business activity.

Whakarewarewa/ Te Puia is the primary geothermal attraction close to the city.

0 km   1 km

(pp. 41–42)

1 Phase 1 was about starting the tourism development process with the establishment of attractions and the facilities and infrastructure to support tourists in the region. Phase 2 was about expanding the level of activity and accommodating large numbers of domestic tourists.

2 Due to increasing numbers of domestic tourists, many motels were developed along Fenton Street to meet their needs as they would drive to Rotorua and need cooking facilities. There was also development of attractions that would suit families like Rainbow Springs and the Agrodome.

3 See answers at right on the diagram.

KEY
- O Core
- Hotels
- Motels
- O Facilities and infrastructure
- Geothermal attractions
- Spa resorts
- Secondary attractions

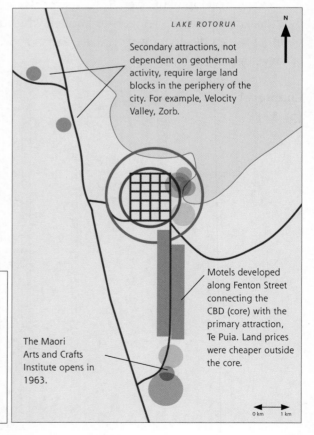

Map showing temporal variations in Phase 2 of tourism development in Rotorua

LAKE ROTORUA

Secondary attractions, not dependent on geothermal activity, require large land blocks in the periphery of the city. For example, Velocity Valley, Zorb.

The Maori Arts and Crafts Institute opens in 1963.

Motels developed along Fenton Street connecting the CBD (core) with the primary attraction, Te Puia. Land prices were cheaper outside the core.

0 km   1 km

(pp. 47–49)

1 Phase 2 was about expanding tourism to meet the needs of increasing domestic tourists. Phase 3 was about greater diversification of activities in the tourism industry to meet the needs of a wider range of tourists.

2 Agglomeration of tourist attractions in clusters makes it more convenient for tourists to visit multiple attractions that might be similar. This also applies to agglomeration of accommodation like hotels and facilities like bars and restaurants in the CBD, which are all centrally located.

3 Global crises can bring change to tourism development by influencing tourist perspectives in source countries, which then affects destination locations. Economic crises mean people would have less money to travel. Issues at the destination might see decreased interest or accessibility issues if there are travel restrictions.

4 See answers at right on the diagram.

5 See timeline below.

KEY
- ◯ Core
- ⬤ Hotels
- ⬤ Motels
- ◯ Facilities and infrastructure
- ⬤ Geothermal attractions
- ⬤ Spa resorts
- ⬤ Secondary attractions

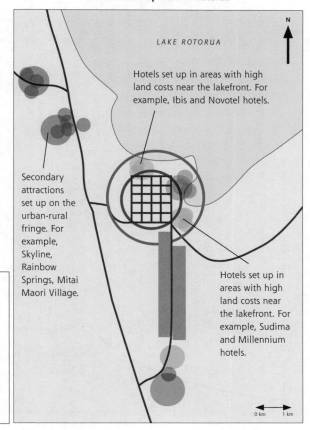

Map showing temporal variations in Phase 3 of tourism development in Rotorua

Hotels set up in areas with high land costs near the lakefront. For example, Ibis and Novotel hotels.

Secondary attractions set up on the urban-rural fringe. For example, Skyline, Rainbow Springs, Mitai Maori Village.

Hotels set up in areas with high land costs near the lakefront. For example, Sudima and Millennium hotels.

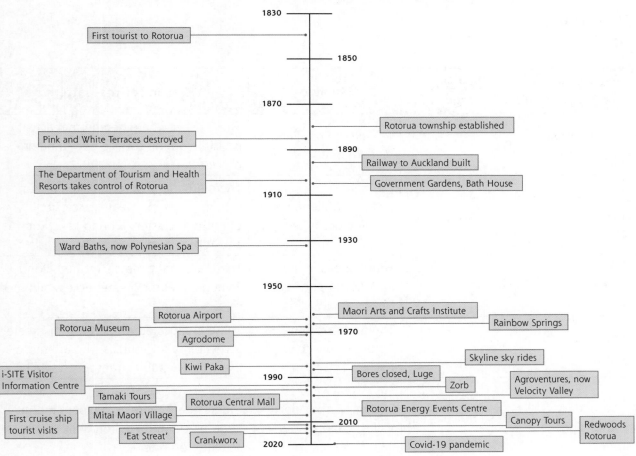

Timeline of key events in the tourism development process in Rotorua (1839–2020)

- 1830
- First tourist to Rotorua
- 1850
- 1870
- Rotorua township established
- Pink and White Terraces destroyed
- 1890
- Railway to Auckland built
- The Department of Tourism and Health Resorts takes control of Rotorua
- Government Gardens, Bath House
- 1910
- Ward Baths, now Polynesian Spa
- 1930
- 1950
- Rotorua Airport
- Maori Arts and Crafts Institute
- Rotorua Museum
- Rainbow Springs
- Agrodome
- 1970
- Skyline sky rides
- Kiwi Paka
- i-SITE Visitor Information Centre
- Bores closed, Luge
- 1990
- Zorb
- Agroventures, now Velocity Valley
- Tamaki Tours
- Rotorua Central Mall
- Rotorua Energy Events Centre
- First cruise ship tourist visits
- Mitai Maori Village
- 2010
- Canopy Tours
- Redwoods Rotorua
- 'Eat Streat'
- Crankworx
- 2020
- Covid-19 pandemic

ISBN: 9780170446914   PHOTOCOPYING OF THIS PAGE IS RESTRICTED UNDER LAW.

(pp. 50–51)

1 Perspectives relate to the changes seen in the tourism development process because tourist perspectives on their needs and wants drives demand for tourism-related activities. At the destination, the tourism industry needs to be aware of tourist perspectives, so their services attract tourists to the area.

2 See answers below on the graph.

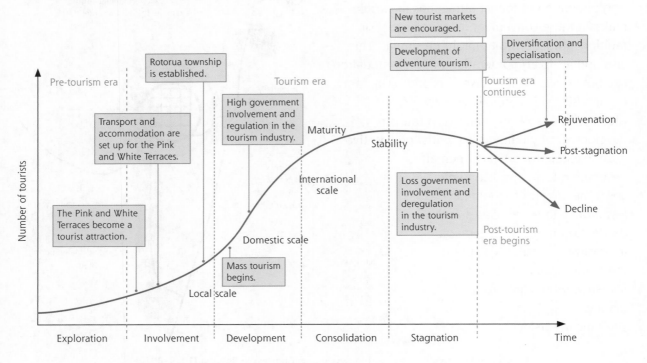

Chapter 6
(p. 53)

Elements/ features	Pattern type	Examples	Reason for the pattern
Geothermal attractions	Dispersed	Te Puia, Kuirau Park, Wai-O-Tapu	These occur where geothermal activity is visible on the surface due to natural processes.
Hotels	Linear, cluster	Sudima, Millennium, Ibis, Rydges hotels	There are many hotels located along Fenton Street for greater accessibility, but also in clusters around Lake Rotorua and Whakarewarewa.
Motels	Linear	Four Canoes, Kuirau Park Motor Lodge, Fenton Court Motel	There are many motels located along Fenton Street for greater accessibility for cars.
Backpackers/ hostels	Cluster	The Backyard Inn, Rotorua Downtown Backpackers	These are located in the CBD for greater accessibility for tourists who are more likely to use public transport or walk.
Secondary attractions	Cluster/dispersed	Skyline, Zorb, Velocity Valley, Canopy Tours	These are located on the urban-rural fringe where there is cheaper land in large blocks and environment for the activities on offer.

 ISBN: 9780170446914

(pp. 59–61)

1 There are different spatial patterns for the different types of activities that take place. Accessibility to activities and services are responsible for the cluster patterns that occur, while transportation creates linear patterns.

2 Tourism-related activities have to meet the needs and wants of tourists so they must be adapted to serve those needs. The location of activities and services are chosen to maximise tourist spending.

3 Patterns relate to features of tourism development in Rotorua because they are the result of interactions in the tourism development process. The patterns seen are a response to feedback over time because successful businesses will thrive and lead to cumulative causation and agglomeration. This creates the spatial patterns seen in the environment.

4

Time period	Pattern type	Link to tourism development	Explanation of link	Interacting elements
Phase 1	**Dispersed** (Geothermal attractions)	**Butler's model — development** **Awareness**	Tourism development begins with awareness of attractions. Only the most popular geothermal features will become attractions.	Attractions, tourists, facilities and infrastructure.
	Concentrated (Facilities and infrastructure)	**Cumulative causation** **Supply and demand**	The tourism industry will develop from the success of attractions and the facilities and infrastructure set up by creating more tourist demand.	Attractions, tourists, facilities and infrastructure, tourism industry, regulators.
Phase 2	**Linear** (Hotels/motels)	**Accessibility** **Affluence**	Consolidation occurs because increasing affluence leads to greater numbers of domestic tourists who have good accessibility to the region.	Tourists, facilities and infrastructure, tourism industry.
Phase 3	**Concentrated** (Backpackers/ hostels)	**Accessibility** **Supply and demand**	After a period of stagnation, new markets were targeted so accommodation types were developed to meet their needs.	Tourists, facilities and infrastructure, tourism industry.
	Clusters (Secondary attractions)	**Butler's model – rejuvenation** **Agglomeration**	Rejuvenation occurred with the development of adventure tourism and less government regulation. This attracted new tourist markets and new clusters of attractions.	Attractions, tourists, facilities and infrastructure, tourism industry, regulators.

5 See answers at right on the diagram.

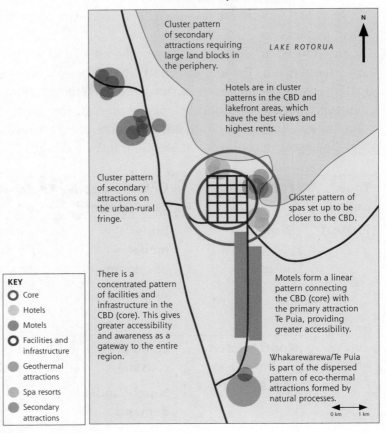

Map showing spatial variations in tourism development in Rotorua

Cluster pattern of secondary attractions requiring large land blocks in the periphery.

LAKE ROTORUA

N

Hotels are in cluster patterns in the CBD and lakefront areas, which have the best views and highest rents.

Cluster pattern of secondary attractions on the urban-rural fringe.

Cluster pattern of spas set up to be closer to the CBD.

KEY
- ⭕ Core
- ⬤ Hotels
- ⬤ Motels
- ⭕ Facilities and infrastructure
- ⬤ Geothermal attractions
- ⬤ Spa resorts
- ⬤ Secondary attractions

There is a concentrated pattern of facilities and infrastructure in the CBD (core). This gives greater accessibility and awareness as a gateway to the entire region.

Motels form a linear pattern connecting the CBD (core) with the primary attraction Te Puia, providing greater accessibility.

Whakarewarewa/Te Puia is part of the dispersed pattern of eco-thermal attractions formed by natural processes.

0 km   1 km

Chapter 7
(pp. 65–67)

1 Student response. A balanced answer would consider the social, economic and environmental effects.

2 The key geographic concept of environments relates to the impacts of the tourism development because tourism development is dependent on the characteristics of both the natural and cultural environments to allow for attractions and a tourism industry to develop. The ongoing tourism development process will modify the environment with a goal to continue tourism development in the future.

3 Sustainability relates to the effects of tourism development on people and the environment and how well tourism-related activities can continue into the future. If there are too many negative impacts, these could cause social, economic or environmental harm and the process would not allow future generations to meet their needs.

     ISBN: 9780170446914

4

Effect	Economic	Social	Environmental
Positive	Jobs and incomes for local people and businesses. Tax revenues for local and central government.	Sharing knowledge and culture with others. Sense of enjoyment and relaxation.	Environmental restoration projects, eco-tourism and conservation attractions.
Negative	Seasonal jobs are temporary and low paid. Costs associated with infrastructure development to accommodate tourists.	Increased visitor numbers could lead to overcrowding. Commercialising culture to make money.	Air pollution from traffic and transportation. Increased waste from tourists. Increased use of ground water could affect geothermal activity.

5 See answers at right on the diagram.

Map showing positive and negative tourism development in Rotorua

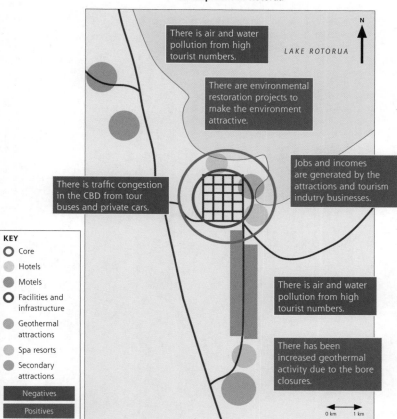